From Design to Discovery

PROFILES, PATHWAYS, AND DREAMS

Autobiographies of Eminent Chemists

Jeffrey I. Seeman, Series Editor

From Design to Discovery

Donald J. Cram

American Chemical Society, Washington, DC 1990

Library of Congress Cataloging-in-Publication Data

Cram, Donald J.
  From design to discovery.

  (Profiles, pathways, and dreams, 1047–8329)
  Includes bibliographical references and index.

  1. Cram, Donald J. 2. Chemists—United States—
Biography. 3. Chemistry, Organic—United States—
History—20th century.

  I. Title. II. Series.

QD22.C7A3 1990 540'.92 [B]   90–607
ISBN 0–8412–1768–8 (cloth)
ISBN 0–8412–1794–7 (pbk.)

The paper used in this publication meets the minimum requirements of American National Standard for Information Sciences—Permanence of Paper for Printed Library Materials, ANSI Z39.48–1984.

PRINTED IN THE UNITED STATES OF AMERICA

Profiles, Pathways, and Dreams

Jeffrey I. Seeman, *Series Editor*
Philip Morris Research Center

M. Joan Comstock, *Head, ACS Books Department*

1990 ACS Books Advisory Board

Foreword

In 1986, the ACS Books Department accepted for publication a collection of autobiographies of organic chemists, to be published in a single volume. However, the authors were much more prolific than the project's editor, Jeffrey I. Seeman, had anticipated, and under his guidance and encouragement, the project took on a life of its own. The original volume evolved into 22 volumes, and the first volume of *Profiles, Pathways, and Dreams: Autobiographies of Eminent Chemists* was published in 1990. Unlike the original volume, the series was structured to include chemical scientists in all specialties, not just organic chemistry. Our hope is that those who know the authors will be confirmed in their admiration for them, and that those who do not know them will find these eminent scientists a source of inspiration and encouragement, not only in any scientific endeavors, but also in life.

M. Joan Comstock
Head, Books Department
American Chemical Society

Contributors

We thank the following corporations and Herchel Smith for their generous financial support of the series Profiles, Pathways, and Dreams.

Akzo nv

Bachem Inc.

E. I. du Pont de Nemours and
Company

Duphar B.V.

Eisai Co., Ltd.

Fujisawa Pharmaceutical Co., Ltd.

Hoechst Celanese Corporation

Imperial Chemical Industries PLC

Kao Corporation

Mitsui Petrochemical Industries,
Ltd.

The NutraSweet Company

Organon International B.V.

Pergamon Press PLC

Pfizer Inc.

Philip Morris

Quest International

Sandoz Pharmaceuticals
Corporation

Sankyo Company, Ltd.

Schering–Plough Corporation

Shionogi Research Laboratories,
Shionogi & Co., Ltd.

Herchel Smith

Suntory Institute for Bioorganic
Research

Takasago International
Corporation

Takeda Chemical Industries, Ltd.

Unilever Research U.S., Inc.

About the Editor

JEFFREY I. SEEMAN received his B.S. with high honors in 1967 from the Stevens Institute of Technology in Hoboken, New Jersey, and his Ph.D. in organic chemistry in 1971 from the University of California, Berkeley. Following a two-year staff fellowship at the Laboratory of Chemical Physics of the National Institutes of Health in Bethesda, Maryland, he joined the Philip Morris Research Center in Richmond, Virginia, where he is currently a senior scientist and project leader. In 1983–1984, he enjoyed a sabbatical year at the Dyson Perrins Laboratory in Oxford, England, and claims to have visited more than 90% of the castles in England, Wales, and Scotland.

Seeman's 80 published papers include research in the areas of photochemistry, nicotine and tobacco alkaloid chemistry and synthesis, conformational analysis, pyrolysis chemistry, organotransition metal chemistry, the use of cyclodextrins for chiral recognition, and structure–activity relationships in olfaction. He was a plenary lecturer at the Eighth IUPAC Conference on Physical Organic Chemistry held in Tokyo in 1986 and has been an invited lecturer at numerous scientific meetings and universities. Currently, Seeman serves on the Petroleum Research Fund Advisory Board. He continues to count Nero Wolfe and Archie Goodwin among his best friends.

Contents

Photographs

Preface

"HOW DID YOU GET THE IDEA—and the good fortune—to convince 22 world-famous chemists to write their autobiographies?" This question has been asked of me, in these or similar words, frequently over the past several years. I hope to explain in this preface how the project came about, how the contributors were chosen, what the editorial ground rules were, what was the editorial context in which these scientists wrote their stories, and the answers to related issues. Furthermore, several authors specifically requested that the project's boundary conditions be known.

As I was preparing an article[1] for *Chemical Reviews* on the Curtin–Hammett principle, I became interested in the people who did the work and the human side of the scientific developments. I am a chemist, and I also have a deep appreciation of history, especially in the sense of individual accomplishments. Readers' responses to the historical section of that review encouraged me to take an active interest in the history of chemistry. The concept for Profiles, Pathways, and Dreams resulted from that interest.

My goal for Profiles was to document the development of modern organic chemistry by having individual chemists discuss their roles in this development. Authors were not chosen to represent my choice of the world's "best" organic chemists, as one might choose the "baseball all-star team of the century". Such an attempt would be foolish: Even the selection committees for the Nobel prizes do not make their decisions on such a premise.

The selection criteria were numerous. Each individual had to have made seminal contributions to organic chemistry over a multidecade career. (The average age of the authors is over 70!) Profiles would represent scientists born and professionally productive in different countries. (Chemistry in 13 countries is detailed.) Taken together, these individuals were to have conducted research in nearly all subspecialties of organic chemistry. Invitations to contribute were based on solicited advice and on recommendations of chemists from five continents, including nearly all of the contributors. The final assemblage was selected entirely and exclusively by me. Not all who were invited chose to participate, and not all who should have been invited could be asked.

A very detailed four-page document was sent to the contributors, in which they were informed that the objectives of the series were

1. to delineate the overall scientific development of organic chemistry during the past 30—40 years, a period during which this field has dramatically changed and matured;

2. to describe the development of specific areas of organic chemistry; to highlight the crucial discoveries and to examine the impact they have had on the continuing development in the field;

3. to focus attention on the research of some of the seminal contributors to organic chemistry; to indicate how their research programs progressed over a 20—40-year period; and

4. to provide a documented source for individuals interested in the hows and whys of the development of modern organic chemistry.

One noted scientist explained his refusal to contribute a volume by saying, in part, that "it is extraordinarily difficult to write in good taste about oneself. Only if one can manage a humorous and light touch does it come off well. Naturally, I would like to place my work in what I consider its true scientific perspective, but . . ."

Each autobiography reflects the author's science, his lifestyle, and the style of his research. Naturally, the volumes are not uniform, although each author attempted to follow the guidelines. "To write in good taste" was not an objective of the series. On the contrary, the authors were specifically requested not to write a review article of their field, but to detail their own research accomplishments. To the extent that this instruction was followed and the result is not "in good taste", then these are criticisms that I, as editor, must bear, not the writer.

As in any project, I have a few regrets. It is truly sad that Egbert Havinga, who wrote one volume, and David Ginsburg, who translated another, died during the development of this project. There have been many rewards, some of which are documented in my personal account of this project, entitled "Extracting the Essence: Adventures of an Editor" published in CHEMTECH.[2]

Acknowledgments

I join the entire chemical community in offering each author unbounded thanks. I thank their families and their secretaries for their contributions. Furthermore, I thank numerous chemists for reading and reviewing the chapters, for lending photographs, for sharing information, and for providing each of the authors and me the encouragement to proceed in a project that was far more costly in time and energy than any of us had anticipated.

I thank my employer, Philip Morris USA, and J. Charles, R. N. Ferguson, K. Houghton, and W. F. Kuhn, for without their support, Profiles, Pathways, and Dreams could not have been. I thank ACS Books, and in particular, Robin Giroux (acquisitions editor), Karen Schools Colson (production manager), Janet Dodd (senior editor), Joan Comstock (department head), and their staff for their hard work, dedication, and support. Each reader no doubt joins me in thanking 24 corporations and Herchel Smith for financial support for the project.

I thank my wife Suzanne, for she assisted Profiles in both practical and emotional ways. I thank my children Jonathan and Brooke for their patient support and understanding; remarkably, I have been working on Profiles for more than half of their lives—probably the only half that they can remember! My family hardly knows a husband or father who doesn't live the life of an editor. Finally, I again thank all those mentioned and especially my family, friends, colleagues, and the 22 authors for allowing me to share this experience with them.

JEFFREY I. SEEMAN
Philip Morris Research Center
Richmond, VA 23234

February 15, 1990

[1] Seeman, J. I. *Chem. Rev.* **1983**, *83*, 83–134.
[2] Seeman, J. I. *CHEMTECH* **1990**, *20*(2), 86–90.

Editor's Note

I REMEMBER STUDYING, as an undergraduate in 1965, Cram and Hammond's *Organic Chemistry*, second edition, in which the front end paper depicted 10 compounds that had been synthesized since the publication of the first edition [including benzyne, barralene, quadricyclane, and housane (bicyclo[2.1.0]pentane)], and the back end paper showed 12 other compounds that, at the time of publication, had not yet been synthesized (including cubane, prismane, and bicyclo[1.1.0]butane). I remember pulling this treasured text off the shelf and placing a mark by some of these compounds as their preparations were reported. That exercise helped me first recognize and then better understand the role of challenge and response in organic chemistry.

> Challenge has been my life. Challenge and response,
> time after time. People should be stretching always, in
> my book, for something a little better.

These few words of Donald J. Cram accurately summarize his philosophy of life. Cram's life provides much evidence of where this philosophy stems from and how it is manifested.

In his youth, Cram overcame a series of obstacles and met varied challenges. His father died when Cram was 4 years old, and the family was quite poor financially. Cram became a barterer, trading his time and energy for food, clothes, and even dental work. Odd jobs and varied chores paid his way through a private high school and led to a college scholarship. There is little doubt that these experiences reinforced a self-reliant, purposeful, and vigorous way of life.

One of Cram's long-time friends and associates describes him as being "monolithic", which, according to Webster, is "an organized whole that acts as a single unified powerful or influential force." Cram's approach to his Profiles manuscript can be understood in this context. He was one of the first authors to submit a first draft, some 62 handwritten pages, and an outline of the remainder on January 14, 1987. "I anticipate I am 55% through, and that my handwritten copy will come to about 120 pages .... Are there too few or too many anecdotes? Is it too detailed, or not detailed enough?" I immediately assured him that his draft was excellent. Within 4 months, I received a completed contribution. Not one other author was so prompt, so determined, and so precise.

And so intense. We are all fortunate, for had Cram been a little less prompt, we might not have received his autobiography at all. Within months of completing his manuscript, Cram had received the Nobel prize in chemistry; whatever discretionary time he had immediately disappeared following that November 1987 announcement.

Interestingly, Cram's autobiography can be divided into two parts, in terms of content, context, and chronology. The first part is sprinkled with anecdotes about and insight into the author. The second part is a very serious description of host–guest chemistry. This writing style echoes the focus and intensity of his scientific life style. Cram can reflect on his pre-1970 years with a greater sense of detachment and humor, while treating the host–guest studies with more seriousness—for they are ongoing!

Cram is an exceedingly hard-working individual, and he creates an environment conducive to that intent. He has been almost a hermit for many years—hard to get out of Los Angeles—as he places his maximum energies on his science. June Hendrix, his secretary for 14 years, is a veritable Praetorian Guard, shielding him from the inconveniences of the outside world, taking messages and relaying responses, thereby allowing Cram to put his full concentration on science. Recognizing this situation, I soon decided which communications and problems could be handled via June (most were!) and which required Cram himself.

Cram was also the first and only author to *demand* that his structures be drawn by the artist of his choice. In fact, most authors were delighted that ACS Books had agreed to my request that the authors' hand-drawn structures be converted to camera-ready format by the ACS. "I trust no one but June Hendrix, my secretary and graphics assistant, to do [the structures] properly . . . so the chapter is uniform and we can control their quality."

Cram's scientific intensity is matched by his serious attitude about and concern for his students. Past students of his applaud his interest and friendship toward them. He is kind to them and appreciative of their contributions to the work. One of Cram's Ph.D. students related the joy and enthusiasm he felt recently when hearing Cram present a plenary lecture: "It was the first time in years that I had that I missed the lab. . . . I have not had that feeling since I moved to law 20 years ago."

Cram has a passion for sports, particularly for those that require physical courage. "He has a 'devil-may-care' attitude on sports, a holy terror!" concludes a Cram friend. Perhaps also a prideful maverick attitude, as Cram pursues surfing and snow skiing. The search for the "big wave" began when Cram was 39. "I like the fear of the big wave . . . to catch the big wave and to be scared to death while being able to handle it."

From Design to Discovery

Donald J. Cram

Donald J. Cram

Personal Notes

An abiding interest in organic chemical research that has lasted for over half a century must have strong personal driving forces. As a boy with a semirural Vermont upbringing (supported by the Aid to Dependent Children program), I had had, by the age of 16, 18 different remunerative, repetitive, and tedious jobs whose tasks were set by others. At that time (1935), my family dispersed, and in succession I spent a year in Florida and a year on Long Island attending high schools and working to support myself. In 1937, I won a four-year scholarship to Rollins College, where I spent four halcyon years. When I first heard the word

Donald J. Cram showing off even at the tender age of five years in Brattleboro, Vermont, 1924.

3

Joanna Cram and her five children, from left to right, Kathleen, Margaret, Joey, Elizabeth, and Donald, in Brattleboro, Vermont, about 1928. Notice how pleased Donald looks at having his picture taken.

research defined by Dr. Guy Waddington at Rollins College in 1937, it had the ring in my mind's ear of variety, expanse, challenge, vitality, and self-determination. The structures of organic compounds, when I first encountered them (1938), engaged and excited my mind's eye. A course in organic chemistry taught by Dr. Eugene Farley in 1938 revealed that, with a few simple rules, new organic structures could be designed. Compounds having these structures were potentially subject to rational synthesis. By the time the second course in organic chemistry was finished, I learned that the physical, chemical, and biological

Donald J. Cram at the age of 17; photo taken on Long Island in 1936.

Donald J. Cram, standing, fourth from left, high school group graduation picture, 1937, Long Island.

properties of organic compounds could also be designed through the agreeable process of reasoning by analogy. The field took on broad dimensions. Each new structure, new reaction, new synthetic sequence, or new set of predicted properties that I envisioned possessed the potential for success or failure. That research had the appealing luster of reasoned gambling for big stakes was explicitly clear to me at the beginning of my career. The carrot of success and the stick of failure hovered over each research venture.

After graduating from Rollins College in 1941, I attended the University of Nebraska, where in 1942 I obtained an M.S. degree. My thesis work was thoroughly enjoyable and was carried out under the direction of Norman H. Cromwell. Because of our entry into World War II, I joined Merck in 1942, where I worked on the penicillin project.

The question of an academic versus an industrial career was settled by this three-year position in this industry (1942–1945), in which, working under Max Tishler at Merck, I discovered the satisfactions of becoming a craftsman in the laboratory and of reducing to practice what had previously been to me largely a paper science. The additional challenges of potential utility and profit further enriched the subject. I chose the academic side of research, not to avoid applications but to skirt the limitations of mandated searches for applications. The posing and answering of intriguing scientific questions of my own conception became my sovereign goal. The teaching of research methodology to a research group became the envisioned means of reaching that goal. The pressures on a university professor to remain a student throughout a career were most welcome. The felicity of campus life, the spirit of inquiry, the contacts with fresh minds, and the challenges of melding good science into good Ph.D. theses appeared congenial. Accordingly, I left Merck to continue my education.

As a graduate student (Harvard, 1945–1947, Ph.D. degree research done under Louis Fieser), I paged backward in the English and American literature looking for research ideas. I examined the initial publications of my research heroes and was comforted to find that many of the early efforts of these scientists ranged from modest to sloppy. Some of them appeared incapable of doing research themselves but were excellent in directing that of co-workers. The tracing of research themes showed how dependent each generation of investigators was on preceding generations, and how interwoven and international was the character of research.

At Harvard, I met John D. Roberts, and we became good friends. Because I had lived in Vermont, Florida, Long Island, New York City, Nebraska, and Massachusetts, I wanted to become acquainted with the west coast before selecting a permanent place to live and work. The school I sought had to be the best place on the west coast for research in organic chemistry and also close to the ocean. Jack described the differences between California Institute of Technology (Caltech), The University of Southern California (USC), and The University of California, Los Angeles (UCLA) and suggested that I go to UCLA. He warmly recommended me to UCLA, and after doing postdoctoral research for him (at MIT) for three months (May, June, and July 1947) to earn money to get there, I was on my way to California. I liked UCLA so much that the rest of my career was spent there.

At the outset of my career in 1941, I was inspired by the wealth of different phenomena encountered in organic chemistry. Research in this science appeared to be a vast playground on which many new games could be invented, games whose players might engage in civil and international competition to discover the appropriate governing rules. What was inconceivable at that time was how the science was to grow in the following decades, and how exciting it would be to contribute to that growth. The wonder at the vast scientific content of organic chemistry with which I started my research has, with time, turned into awe and pride. The field not only grew, but it met and enveloped contiguous fields of research such as genetics and biology. Historians may look back on the second half of the 20th century as the golden age of the science of organic chemistry.

I have always regarded scientific endeavor as highly personal. The chemistry of the day, to get done, had to be driven by the thought that it is the most important thing in the world, but the chemistry of the paper, thesis, or monograph requires perspective, context, and balanced judgments. The following pages illustrate my inability to fully make the transition between these extremes. To be written at all, these pages had to become the "chemistry of the day."

From Mold Metabolites to Phenonium Ions (1947–1969)

By 1947, exploration of the chemical structures of the natural products of the microbiological world had revealed the presence of functional groups, configurations of asymmetric centers, and ring systems not found in other natural products. My first research problem at UCLA as an instructor and independent investigator involved the isolation and determination of the structure of sorbicillin[1] (1), a by-product of penicillin production at Merck.[2] The second problem involved the structure of 2, an optically active degradation product of the antibiotic citrinin. I observed that when 2 was heated in 4 N aqueous sulfuric acid, the compound racemized.[3] This result led directly to a detailed mechanistic study of the Wagner–Meerwein rearrangement that employed, for that

1   **2**

time (1948), an unusual feature. Simultaneous configurational changes at two chiral centers during an aryl migration from one to the other were used as structural probes for determination of the symmetry properties of short-lived cationic reaction intermediates.

I based this study[3] on the generalization that the ratio of products derived from a particular reaction intermediate was independent of the starting material that produced that intermediate. All four enantiomers of **3** were prepared (Ts is p-CH$_3$C$_6$H$_4$SO$_2$), and their relative configurations were determined. When solvolyzed in AcOH (Ac is CH$_3$CO), each enantiomer of *threo*-**3**-OTs gave racemic *threo*-**3**-OAc, but no racemic *erythro*-**3**-OAc. Each enantiomer of *erythro*-**3**-OTs gave only *erythro*-**3**-OAc, without configurational modification. These results suggested the intervention of the symmetrical *cis*-phenonium ion in the *threo*-series and of the asymmetric *trans*-phenonium ion in the *erythro*-series. Similar studies[4,5] of the stereoisomers of systems **4** and **5** gave the results expected on the basis of the intervention of the appropriate diastereomeric *cis*- and *trans*-phenonium ion intermediates in this system as well. This experiment revealed the rearrangement that had been invisible in the experiment with *erythro*-**3**-OTs.

In a second study[6] (1952), I found that superimposed on the acetolysis was an ion-pair collapse process, termed by my colleague Saul Winstein, *internal return*.[7] Thus, (−)-*threo*-**3**-OTs was observed to give racemic *threo*-**3**-OTs in a process competitive with acetolysis. Both processes involved the *cis*-phenonium tosylate ion pair as a short-lived reaction intermediate. As expected, (+)-*erythro*-**3**-OTs underwent an invisible rearrangement to itself through the *trans*-phenonium tosylate ion pair. This reaction was revealed by introduction of p-BrC$_6$H$_4$SO$_3^-$ (BsO$^-$) into the medium. In a reaction competitive with solvolysis and collapse, the ion pair underwent anion exchange and collapse to give (+)-*erythro*-**3**-OBs mixed with (+)-*erythro*-**3**-OTs.

This work was done with my own hands and provided that sense of intimacy and identification with research results that is impossible to achieve fully when co-workers do the experiments. I take pride in the fact that I earned tenure with my personal research. Tenure was

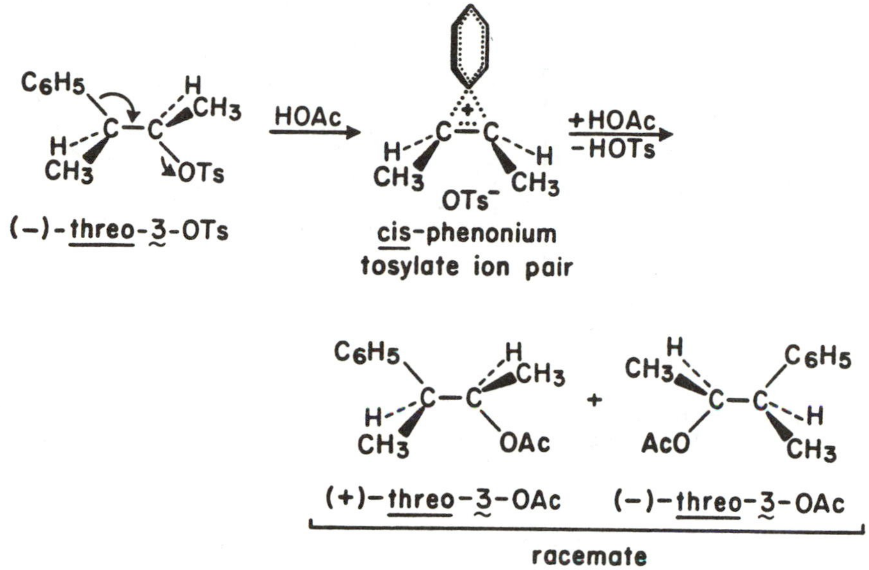

HOAc
+HOAc
-HOTs
(−)-threo-3-OTs
cis-phenonium
tosylate ion pair
(+)-threo-3-OAc
(−)-threo-3-OAc
racemate

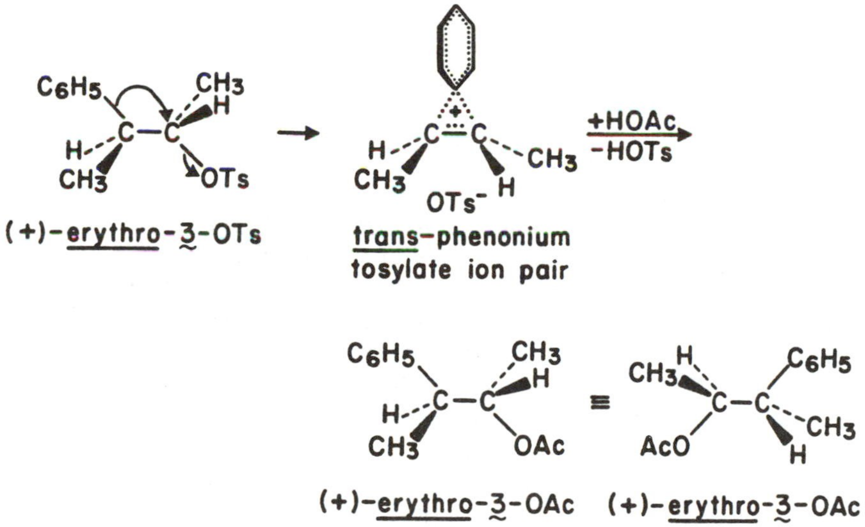

+HOAc
-HOTs
(+)-erythro-3-OTs
trans-phenonium
tosylate ion pair
(+)-erythro-3-OAc
(+)-erythro-3-OAc

too important to leave to graduate students. In those days, my students could talk to me only while I shook separatory funnels and took melting points.

My 1948–1949 work preceded by several years that of Winstein et al.,[7,8] who made somewhat similar use of stereochemical probes of the mechanism of the Wagner–Meerwein rearrangement in the norbornyl system. However, Winstein's discovery of internal return[7] in the nor-

```
        OBs
         |
CH₃—CH—CH—CH₂—CH₃ ⎤
     |                ⎥
    C₆H₅             ⎥
                     ⎥
Single stereo-       ⎥    cis- or trans-   -HOBs
isomers of 4-OBs   ⎯→ ⎥    phenonium        ⎯⎯⎯⎯→
                     ⎥    ion pairs        +AcOH
        OBs          ⎥
         |           ⎥
CH₃—CH₂—CH—CH—CH₃ ⎦
            |
           C₆H₅

Single stereo-
isomers of 5-OBs
```

$$CH_3—CH—\underset{|}{CH}—CH_2—CH_3 \;+\; CH_3—CH_2—\underset{|}{CH}—CH—CH_3$$

```
        OAc                              OAc
         |                                |
CH₃—CH—CH—CH₂—CH₃  +  CH₃—CH₂—CH—CH—CH₃
     |                                    |
    C₆H₅                                 C₆H₅


      Single stereo-              Single stereo-
      isomer of 4-OAc            isomer of 5-OAc
```

bornyl system preceded by about a year our detection of the same phenomenon in the open-chain 3-phenyl-2-butyl system.[6]

The competition between Winstein and me provided a wholesome stimulation for both of us that was punctuated occasionally by intense arguments. I acknowledge my debt to him. He was the most thorough and tenacious critic of research results I have encountered. His stock in trade reduced to the question: "What do you really know, and how do you know it?" He was generous with criticism and stingy with praise, traits that, given our respective characters, were much better for my career than if the opposite had been true. What I really admired about him was his devotion to the science, to inquiry, and to truth.

Seminars given at UCLA in the late 1950s were real adventures that were enriched by the attendance of John D. Roberts and George S. Hammond. I am indebted to Roberts for his suggestion that I go to UCLA and for his letter of recommendation; I am indebted to Hammond for accompanying me there and for many fruitful discussions about research and teaching over the years.

At UCLA, circa 1959. Left to right, Donald J. Cram, Kenneth Conrow, G. Ross Robertson, S. Winstein, E. R. H. Jones, Tom Jacobs, Jim Hendrickson, and Clifford Bunton. Photo lent by Carolee Winstein.

Saul Winstein and Donald J. Cram, circa 1962. Photo taken by R. Huisgen at UCLA.

My phenonium ion work stimulated much research and generated much controversy, but fortunately, much less of the latter than Winstein's norbornyl nonclassical ion results. Both my original experimental results and interpretations withstood the test of time, many new experiments, and new instrumentation. A definitive review has appeared.[9]

Cram's Rule (1952–1963)

In the course of synthesizing the stereoisomers and determining the configurations of compounds of the type C^*-C^*-OH (asterisks identify asymmetric centers), we encountered many examples of the generalizations of McKenzie and Tiffeneau. We summarized them as follows: "In the creation of a new asymmetric center in the presence of an old one, the configuration at the new asymmetric center of the predominant diastereomeric product can be inverted by inverting the order in which substituents are attached to that new center."[10]

For our phenonium ion studies, we had to establish the relative configurations of the two asymmetric carbons of our products. As a result, by 1952, we had investigated nine examples of conversions of general compound 6 to predominant diastereomer 7. These results led

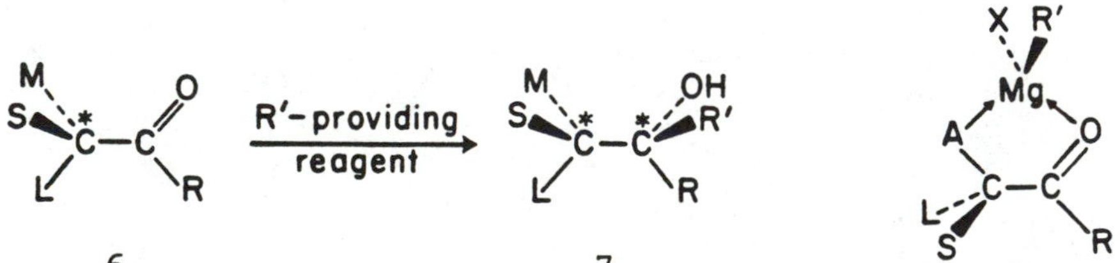

us to formulate the rule of *steric control of asymmetric induction* in reactions of acyclic systems in which a new asymmetric center is created adjacent to an old:

> In noncatalytic reactions of the type shown, that diastereomer will predominate which would be formed by the approach of the entering group from the least hindered side of the double bond when the rotational conformation of the C—C bond is such that the double bond is flanked by the two least bulky groups attached to the adjacent asymmetric center.[10]

In 6 and 7, L, M, and S symbolize large, medium, and small substituents, respectively. Our nine examples, combined with those found in the literature, successfully correlated the configurations of 35 compounds prepared by six different reactions of the type just described. We then used the rule to predict the configurations of 50 compounds found in the literature whose stereochemical structures had hitherto not been assigned.[10]

In the original 1952 paper we stated that

> the rationale for operation of the rule may be quite different for different cases. [In some cases,] the asymmetric carbon atom of the starting material carries a hydroxyl or amino group which can possibly react with the reagent (e.g., a Grignard) to form a complex . . . . This complex would have the geometry of a five-membered ring, and R' would be expected to approach the carbon atom carrying the carbonyl group from the least-hindered side.[10]

Interestingly, our Communication stating our rule to the Editor of the *Journal of the American Chemical Society* was rejected, but a full-length paper was recommended by the referees and editor. Several months after the full paper had been submitted, Vladimir Prelog gave a seminar at UCLA on Prelog's rule, which dealt with 1,4-asymmetric induction in the reactions with organometallic reagents of α-keto esters of natural-product alcohols containing asymmetric centers. The same kind of thought processes had led Professor Prelog and me to our respective generalizations. As a result of this and many subsequent associations, Vlado Prelog and I became good friends who profitably stimulated each other over much of our respective careers. He is the prototype "chirophile." It takes one to recognize one.

Louis Fieser, under whom I had obtained my Ph.D., was the first to refer to our generalization as "Cram's rule," a name I am little inclined to fault. Hundreds of examples of its applicability, and a few exceptions, have appeared with the passage of time. When the term anti-Cram was first applied to the exceptions, my first thought was that the term called attention to my fallibility. My second thought was a recollection that P. T. Barnum had once said, "There is no such thing as bad publicity." My third thought drew an analogy between the terms Cram and anti-Cram, and the terms Markovnikov and anti-Markovnikov. Both sets of terms succinctly state a given result whose meaning is clear to the initiated. This nomenclature probably appears more trivial to the reader than to me. When I attended a meeting in Germany recently, a speaker used the term anti-Cram, and told me afterwards that he hoped I felt honored by its use. I told him my heart was warmed more by its use than my pride was hurt by the failure of the rule.

Elimination Reactions

The E_i Reaction

In connection with the phenonium ion work, I had used the predominantly *cis* stereochemical course of the Chugaev reaction (xanthate pyrolysis) to confirm our configurational assignments to the *threo*- and *erythro*-3-phenyl-2-butanols.[11] Huckel et al.[12] had previously assigned such a stereochemical course in the cyclic menthyl and neomenthyl systems. I observed that *threo*-8 gave about a 3 to 1 ratio of alkene 10 to alkene 9, and that *erythro*-8 gave about a 10 to 1 ratio of alkene 9 to 10. This experiment brought the configurational assignments based on the phenonium ion work into conformity with the expected *cis* elimination course of the Chugaev reaction. This work was the first stereochemical study of the E_i (elimination, internal) reaction in an open-chain system.

In 1954, J. E. McCarty and I reported[13] the first investigation of the stereochemical course of the Cope elimination reaction. We determined the configurations of *threo*-11 and *erythro*-11 by independent means. The results of the pyrolysis at 115–125 °C showed the reaction to be more stereoselective than the Chugaev reaction. In a later kinetic study with M. R. V. Sahyun, I observed[14] the enormous effect of solvent on the rates of the reactions. In dry tetrahydrofuran (THF) at 52 °C, the rates were approximately 10^6, and in dry $(CH_3)_2SO$ the rates were approximately 10^5 times faster than in H_2O at the same temperature. The reactions were completely stereospecific at the low temperatures attainable in non-hydrogen-bonding solvents. Solvents capable of hydrogen bonding to the leaving group $((CH_3)_2N^+O^- \cdots HOR)$ radically depressed the activity of the negatively charged oxygen.[14]

By analogy with the amine oxide elimination reaction, we reasoned that pyrolysis of appropriate sulfoxides should also undergo elimination reactions, and that the reactions should assume a predominantly *cis* steric course. In 1960, C. A. Kingsbury and I[15] prepared all four diastereomeric racemates of 12 (sulfur provides an additional asymmetric center). The elimination reactions went as expected, each of the four isomers giving good yields of alkene with high stereoselectivity. All four reactions that were run in dioxane at 80 °C assumed a dominant *cis* stereochemical course, as anticipated. Two of the reactions were formulated (the configurational assignments at sulfur were not made with

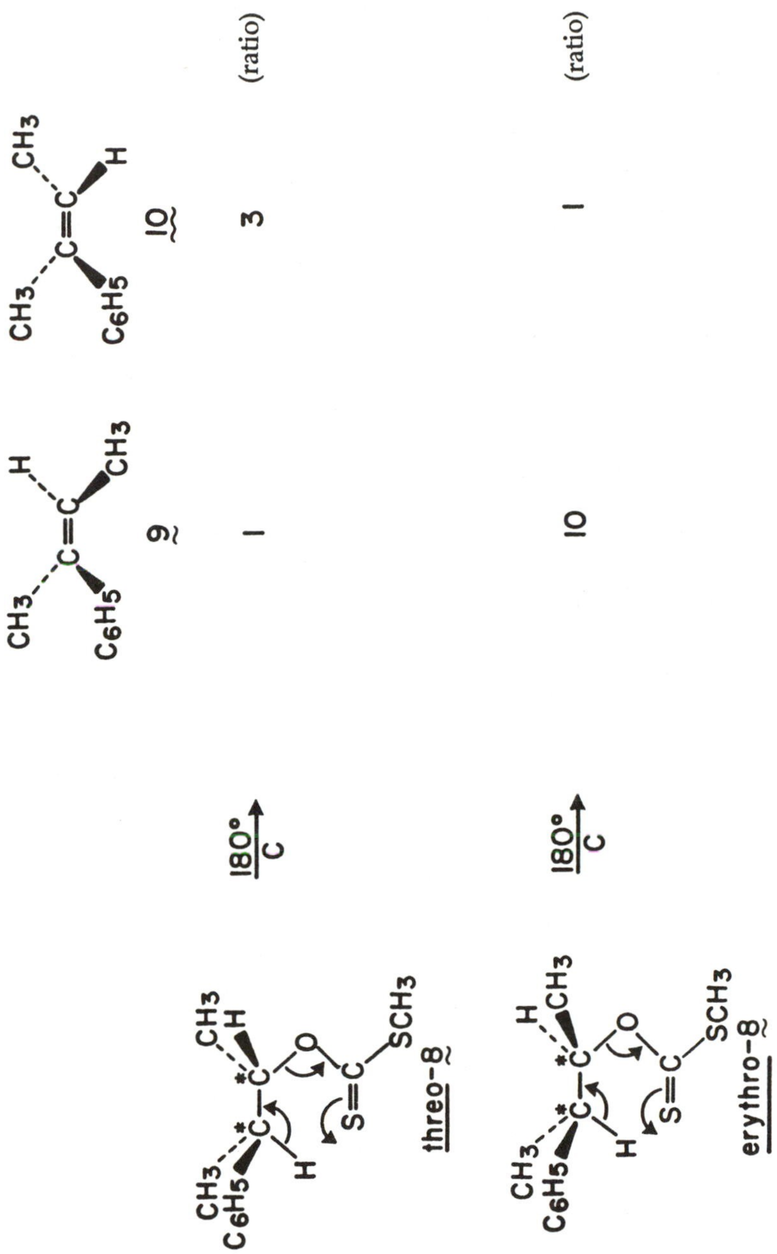

CH$_3$

H

CH$_3$—C=C—C$_6$H$_5$

$\underline{10}$

3

(ratio)

—

(ratio)

H

CH$_3$

CH$_3$—C=C—C$_6$H$_5$

$\underline{9}$

—

10

$\xrightarrow{\underline{180°}\,\text{C}}$

$\xrightarrow{\underline{180°}\,\text{C}}$

CH$_3$ H

C

O

CH$_3$ C$_6$H$_5$

C

H

S=C—SCH$_3$

$\underline{threo\text{-}8}$

H CH$_3$

C

O

CH$_3$ C$_6$H$_5$

C

H

S=C—SCH$_3$

$\underline{erythro\text{-}8}$

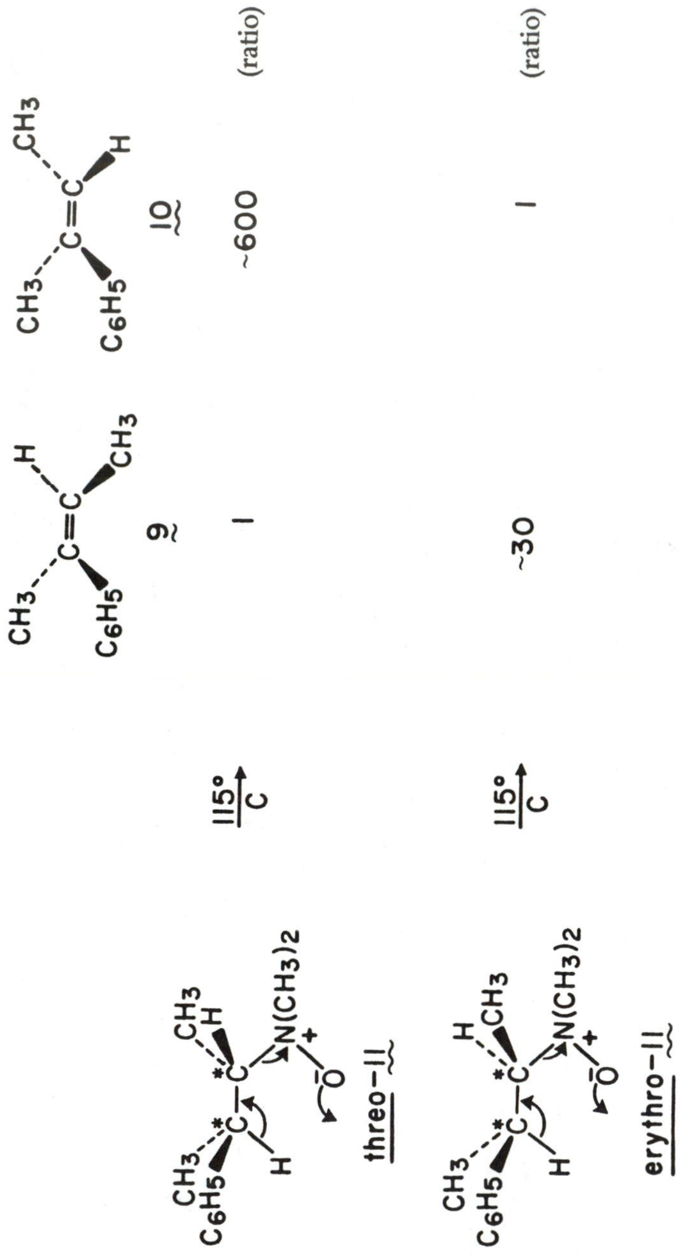
threo-11
erythro-11
115° C
9
10
1
~600
(ratio)
~30
1
(ratio)

certainty). We stated that "because of the relative ease of preparation of sulfoxides and the low temperatures and high yields associated with their transformations to olefin, the reaction may be useful in synthesis." This reaction was the first that we had invented, and we took great pleasure in finding that it assumed the expected stereochemical course. During the 1970s and 1980s, the reaction and its selenium oxide analogue have been applied to synthesis.

The E_2 Reaction

In research done in 1956, F. D. Greene, C. H. DePuy, and I[16] first posed the question of whether eclipsing effects could be used to measure the amount of double-bond character in the transition state of the classical E_2 (second-order elimination) reaction. We formulated the transition state as being composed of three contributing structures, **A**, **B**, and **C**. The relative importance of the three structures should correlate with "the nature of the leaving group, the base strength, and the solvating ability of the solvent." Both diastereomers of the three systems, **15**, **16**, and **17** were designed and prepared. Their bimolecular rate constants to form **13** and **14** were determined by using different basic catalysts of varying strengths. In a model experiment, the diastereomeric formate esters corresponding to **15–17** were equilibrated in formic acid. The *threo*-to-*erythro* ratio at equilibrium was 0.8, a result suggesting that the same should be true for starting materials **15–17**. An equilibration of alkenes **13** and **14** demonstrated that, at equilibrium, **14** dominated **13** by a factor of at least 50 to 1. The bulky phenyl groups are *syn* to one another in **13**, and *anti* in **14**. We reasoned that in the E_2 reaction of **15–17**, high values of $k_{threo}/k_{erythro}$ should reflect high carbon–carbon double bond character in the transition state, as suggested by structure **B**. Values approaching unity should reflect less double bond character, as suggested by structures **A** or **C**.

The results obtained from these highly designed systems were gratifying. As base strength was increased from EtO^- to n-$C_6H_{13}(CH_3)CHO^-$ to t-$C_4H_9O^-$ with Cl as leaving group, $k_{threo}/k_{erythro}$ values went from 1 to 3.5 to 10.7, respectively. We concluded that "this trend probably reflects the amount of assistance that the electron pair of the C–H bond gives to the breaking of the C–X bond." In EtOH–EtONa, in passing from X = Br to Cl to $N^+(CH_3)_3$, the ratio changed from 0.7 to 1.1 to 57, respectively. In solvolyses, reaction rates were Br > Cl > $N^+(CH_3)_3$. Thus, in our E_2 reaction, structure **A** dominates the transition state with Br as a good leaving group, but structure **B** dominates with the poor $N^+(CH_3)_3$ leaving group.

In t-C_4H_9OH– t-C_4H_9OK, in passing from Br to Cl to $(CH_3)_3N^+$, the ratio went respectively from 5.4 to 15 to 1.1. In all reactions except

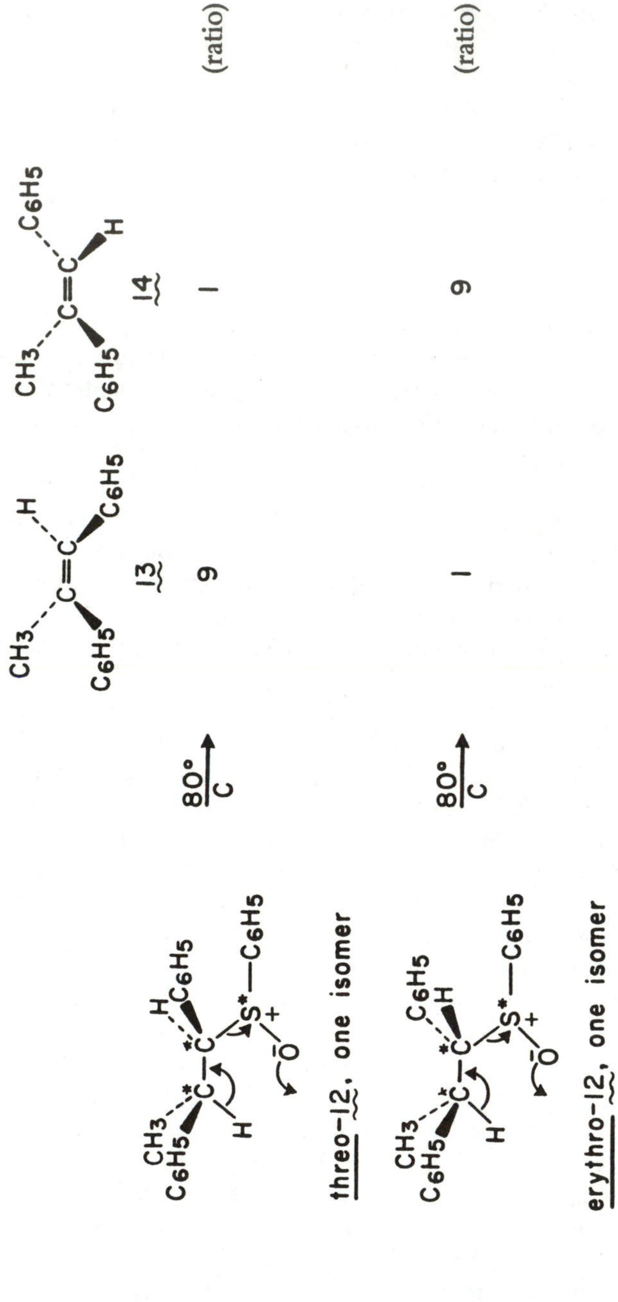

threo-12, one isomer

$\xrightarrow{80°\,C}$

13   9   14   1   (ratio)

erythro-12, one isomer

$\xrightarrow{80°\,C}$

1   9   (ratio)

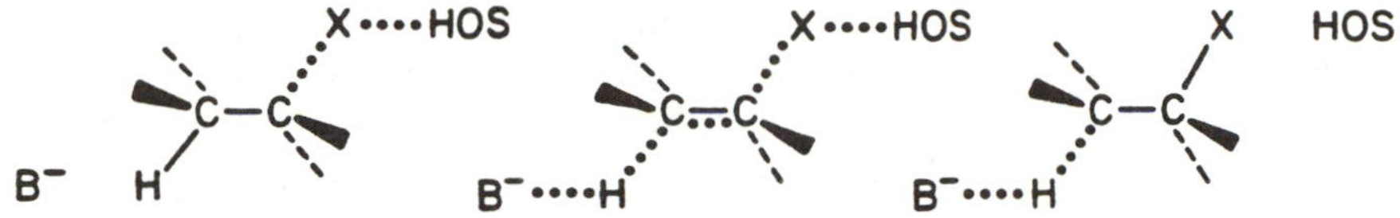

A   B   C

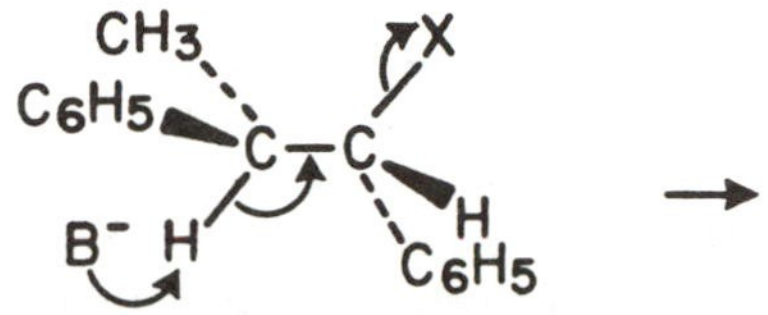

$\underline{\text{threo}} - \underset{\sim}{15}$, X = Cl

$\underline{\text{threo}} - \underset{\sim}{16}$, X = Br

$\underline{\text{threo}} - \underset{\sim}{17}$, X = $\overset{+}{N}(CH_3)_3$

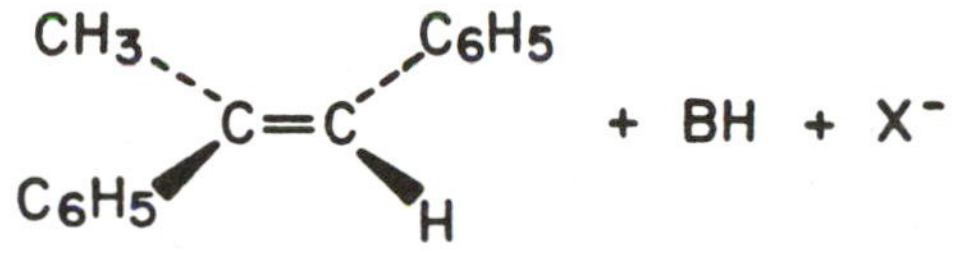

+ BH + X⁻

$\underset{\sim}{14}$

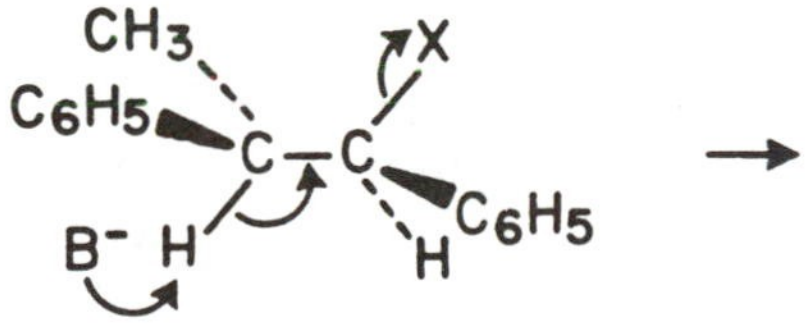

$\underline{\text{erythro}} - \underset{\sim}{15}$, X = Cl

$\underline{\text{erythro}} - \underset{\sim}{16}$, X = Br

$\underline{\text{erythro}} - \underset{\sim}{17}$, X = $\overset{+}{N}(CH_3)_3$

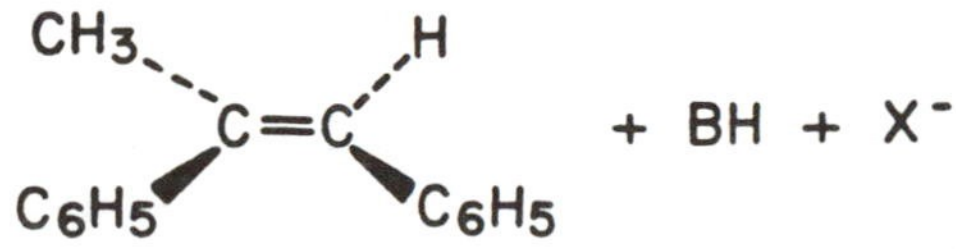

+ BH + X⁻

$\underset{\sim}{13}$

the last one, *threo* starting materials gave only 14 and *erythro* starting materials gave only 13. In the last reaction, only 14 was produced from both diastereomers of 17, a result indicating that *erythro*-17 had epimerized to *threo*-17 faster than the E_2 reaction had occurred. Apparently, the strong base strength of t-C_4H_9OK combined with the acidifying effect of the $(CH_3)_3N^+$ group was enough to generate an α-carbanion intermediate leading to epimerized product.

Our 1956 investigation[16] was the first in a long series of thorough mechanistic studies of the E_2 reaction by C. H. DePuy, W. H. Saunders, Jr., and J. F. Bunnett. Every conceivable technique was applied to this interesting reaction. The use of isotopic and stereochemical techniques proved particularly useful. Our early experiments pointed to the questions to be settled much more than to the answers that ultimately emerged.

The Cyclophanes (1951–1970)

Origins

In 1945, I noted in my idea book that compounds should be prepared and studied in which two aromatic rings are held face-to-face by methylene bridges substituted in their *para* positions. The idea grew out of a reading of Michael Dewar's ideas about possible π-complex intermediates in the benzidine and semidine rearrangements.[17] In the fall of 1948, H. Steinberg, my first graduate student, undertook the syntheses of 18–21. The research was completed in 1950, and published in 1951.[18] We prepared 18 in 2.1% yield by the high-dilution addition of $(4\text{-BrCH}_2C_6H_4CH_2)_2$ in xylene over a 60-h period to refluxing xylene and molten sodium stirred at 7000 rpm.

Meanwhile, a polymer research group at ICI in Great Britain was attempting to prepare and polymerize *p*-xylylene, the tetraene derived by high-temperature cracking of *p*-xylene. Trace amounts of 18 were crystallized from the nonpolymeric fraction. Without the usual elemental analyses and UV spectrum, this curiosity was communicated in 1949 by Brown and Farthing;[19] the compound was characterized only by a later-corrected crystal structure.[20] Steinberg had already prepared 18 by the time we learned of the fortuitous results of the English group. We likewise learned from a literature reexamination that Reichstein and Oppenhauer[21] had attempted to prepare compounds such as 18 in 1933. Furthermore, Baker et al.[22] reported the synthesis of 22 in 1950.

I gave our initial compounds the family name paracyclophanes,[18] but later, pressed by the need to name our growing number of substituted cyclophanes, J. Abell and I[23] broadened and systematized the nomenclature.   Thus, **21** became [3.6]paracyclophane; **22** became [2.2]metacyclophane; and **23** became 4-acetyl[2.2]metaparacyclophane.

Our investigations of phenonium ions and of Cram's rule were serendipitous and evolutionary. One result and idea led to others until patterns emerged and research areas could be mapped. The design of chemical systems to answer scientific questions was a piecemeal proposition in which what was to be done tomorrow depended on what was discovered today. In contrast, the cyclophanes were designed, from the beginning, to answer general scientific questions:

1. How do transannular electronic effects between two or more unsaturated systems depend on their enforced geometric locations with respect to one another? How are such effects transmitted in a variety of chemical circumstances?

2. What are the limits of the structural theory of organic chemistry with respect to deforming and crushing aromatic nuclei?

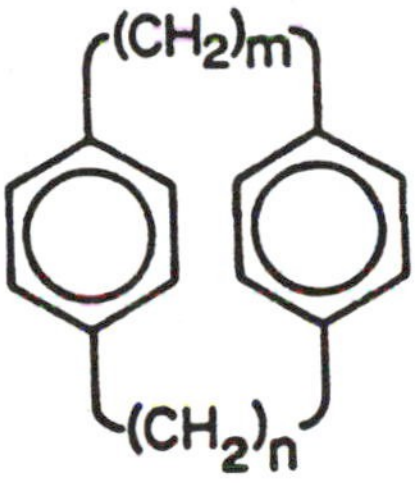

18, m=n=2; 19, m=2, n=3;
20, m=2, n=4; 21, m=3, n=6
or [3.6]paracyclophane

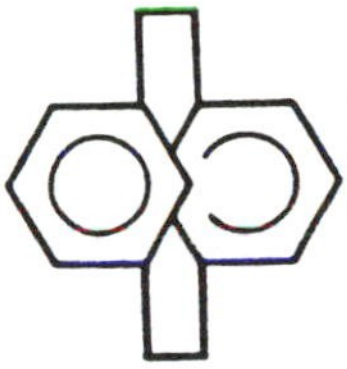

22 or [2.2]meta-
cyclophane

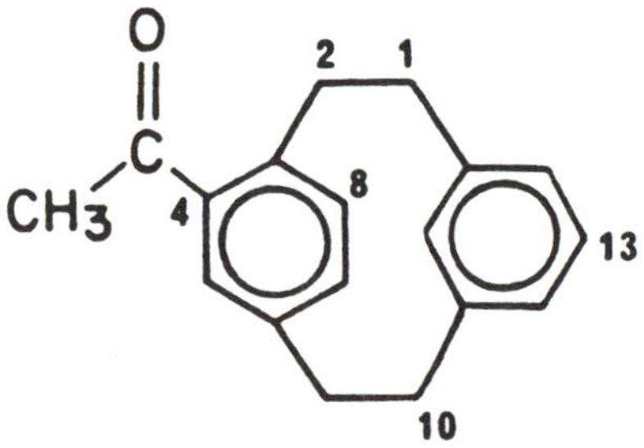

23 or 4-acetyl[2.2]
metaparacyclophane

3. What are the chemical and stereochemical consequences of the presence of great strain in the cyclophanes?

4. What are the predictive powers and limitations of the use of space-occupying molecular models based on crystal structure determinations?

All of these questions were posed, in one form or another, in the first of over 50 full-length papers we published in the cyclophane field. Of course, ancillary questions were posed as time passed, but most of our results fell within the scope of the initial concept.

Syntheses and UV Spectra

The Winberg et al.[24] discovery that [2.2]paracyclophane (18) could be synthesized in good yield from p-$CH_3C_6H_4CH_2N(CH_3)_3OH$ made this compound a more convenient starting material for preparation of the simple cyclophanes originally synthesized by more laborious routes by H. Steinberg[18] and N. L. Allinger.[25] Through ring expansions and rearrangements, R. C. Helgeson,[26] L. A. Singer et al.,[27] M. H. Delton[28], and R. E. Gilman[29] prepared 19 and 24–27.

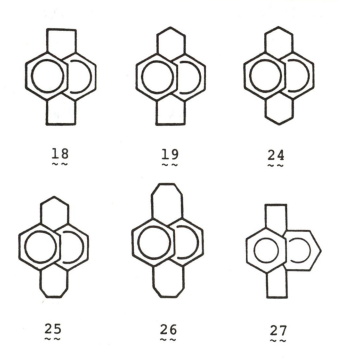

18   19   24

25   26   27

In 1954, N. L. Allinger, H. Steinberg, and I[25] established that the smallest paracyclophane with a normal UV spectrum was 26. The large abnormalities in the smaller homologues were due to bent benzene rings and transannular electronic effects. These two kinds of effects were distinguished through UV spectral comparisons between the [*m.n*]paracyclophanes and the [*m*]paracyclophanes. Of the latter, the smallest homologue that C. S. Montgomery, G. R. Knox, and I[30] prepared had *m* = 8 (29). Compound 29 was synthesized by cycloaddition of *p*-xylylene and its furane analogue, prepared in the same flask by 1,6-elimination–cycloaddition reactions. The product, 28, was converted to 29 by conventional reactions. Allinger[31] calculated that the bridgehead carbons of 29 were bent out of the plane of the other four by 20 °, much more than was observed in the crystal structures of 18[20] or 27.[32] Comparisons of the UV spectra of bent cyclophanes such as 29 that contain a single benzene ring with those containing two bent benzene rings crushed together established the presence of transannular electronic effects in the smaller [*m.n*]paracyclophanes.[33] Much more recently, others synthesized the highly strained 30, without the benzene ring collapsing to give Dewar benzene.[34]

Transannular Electronic Effects

Transannular influences became visible literally through the wide range in color differences (light yellow to deep purple) in the π–π complexes (31) of the homologous [*m.n*]paracyclophanes and tetracyanoethylene (R.

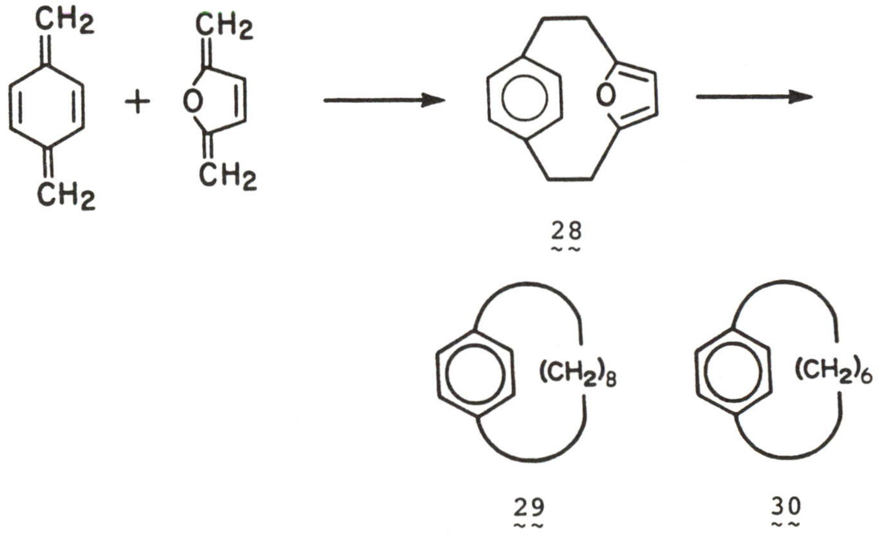

28

29   30

H. Bauer).[35] The π-base strengths decreased in the order [3.3] > [4.3] > [2.2] > [4.4] > [6.6] ~ (p-CH$_3$C$_6$H$_4$CH$_2$)$_2$.

Except for the position of the badly bent [2.2]paracyclophane in this series, the order correlates with the distances of the two benzene rings from one another. The closer the two rings, the greater becomes the π-base strength, the nonbound benzene ring releasing electrons to the bound ring (*see* 31). As expected, electron-releasing substituents in the noncomplexed ring decreased the π basicity of the complexed ring as shown in 32 (L. A. Singer).[36]

31   32

By about 1967, ^{1}H NMR spectral techniques were well-enough developed to allow H. J. Reich[37] to examine transannular directive influences in electrophilic substitution of monosubstituted [2.2]paracyclophanes (e.g., 33). We found that predominant substitution occurred pseudo-*gem* to the most basic positions of substituents in the already-substituted ring, as in the conversion of 33 to 34 or 35 (E. A. Truesdale).[38] That hydrogen was transferred from ring to ring during substitution was demonstrated by both kinetic isotope effects, and the conversion of 36 to 37.[37]

Chemical Effects of Compression

The 31-kcal/mol strain energy in [2.2]paracyclophane[39] is chemically expressed by the cleavage of the compound to form biradicals reversibly at 200 °C. When optically pure 38a was heated to 200 °C, it racemized.[40] The experiments designed to differentiate between such intermediates as 39 and 40 illustrate the use of stereochemical techniques to solve mechanistic problems. Molecular model comparisons indicated there was no possibility of the substituted aryl ring rotating without bonds breaking. Compounds 41 and 43 were prepared and found not to interconvert at temperatures of about 200 °C, conditions at which 41 equilibrated with 42, and 43 equilibrated with 44.[40] Furthermore, when 18 was heated to 200 °C in either dimethyl maleate or fumarate, the system

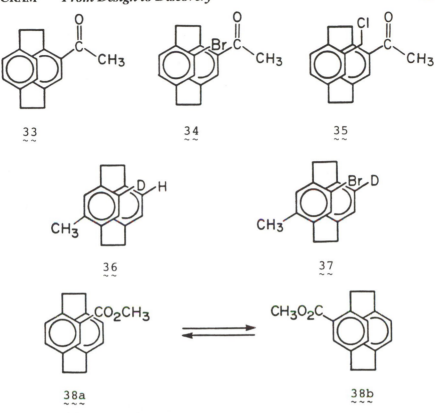

33   34   35

36   37

38a   38b

Donald J. Cram with Jack and Edith Roberts's children, Anne, John Paul, Allen, and Don Roberts, circa 1960. Photo taken by J. D. Roberts.

39a 39b 40

41 42

43 44

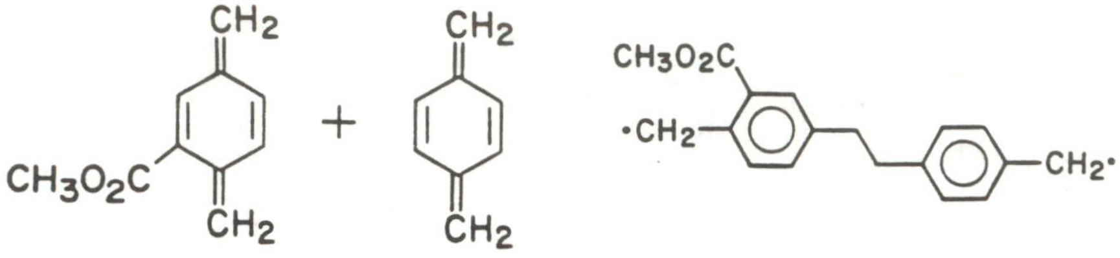

ring expanded to give **45** in a miniature A–B radical polymerization reaction.[40]

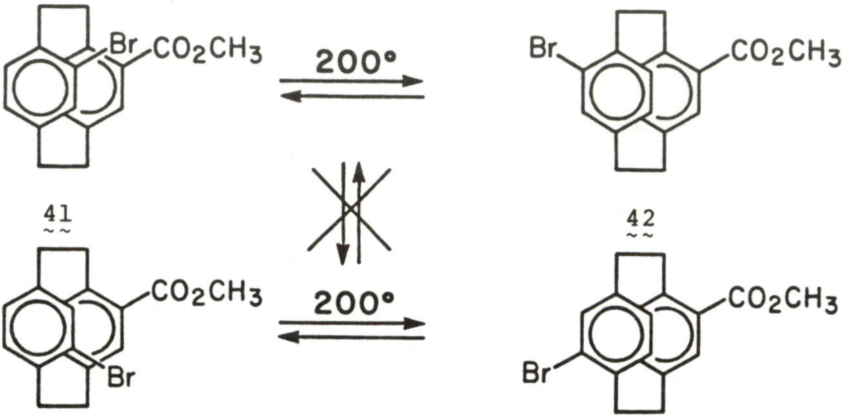

Torsional chirality was observed by enantiomeric resolution of compounds **46–48**; however, in **48**, the substituted benzene ring rotated with respect to the nonsubstituted ring at high temperature at rates on the human time scale of seconds, minutes, hours, and days (W. J. Wechter et al.).[41] Compound **49** resisted resolution into enantiomers, suggesting fast rotations of the benzene rings relative to one another.[41]

18 45 (diastereomeric mixture)

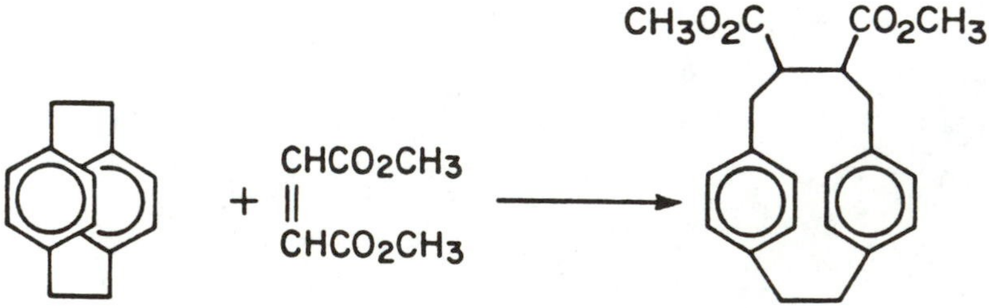

In work done after ^{1}H NMR spectrometers became available, D. T. Hefelfinger[42] and I observed, by a combination of spectral and polarimetric methods, that the *meta*-substituted ring of **50** rotated with respect to the *para* at available temperatures, but not the *para* with respect to the *meta*.

Classically Conjugated Cyclophanes

The synthesis and study of classically conjugated but orbitally unconjugated cyclophanes **51–54** provided K. C. Dewhirst[43] with a thesis in the late 1950s. UV spectral and crystal structural comparisons both showed how unconjugated the unsaturated systems were in **51** and **52**. The multiple bonds of **53** and **54** were only partially conjugated. ^{1}H NMR spectroscopy was not available in 1958, so we could not observe the probable equilibrium between enantiomers **53a** and **53b**. We tried with acid and light at 25 °C to convert **54** to **55**, which at that time would

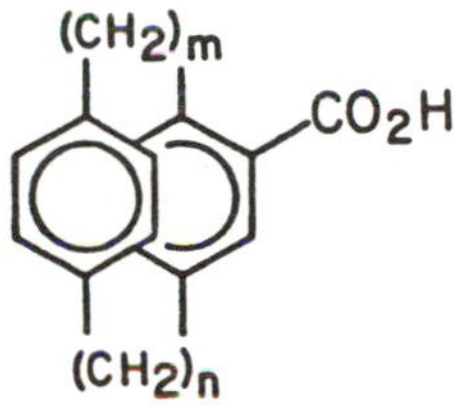

46, m=n=2; 47, m=n=3;
48, m=3, n=4; 49, m=n=4

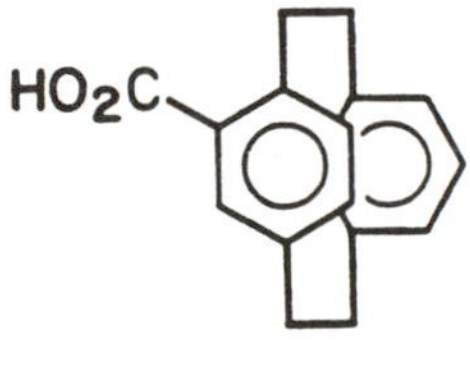

50

have been the first [18]annulene.[44] Almost 20 years later, Boekelheide et al.[45] accomplished this conversion.

The advances in the cyclophane field, since we left it in 1970, were hard to imagine at that time. The international character of the effort is illustrated by a very abbreviated list of the members of the "cyclophane club": V. Boekelheide, P. M. Keehn, S. Misumi, K. Koga, I.

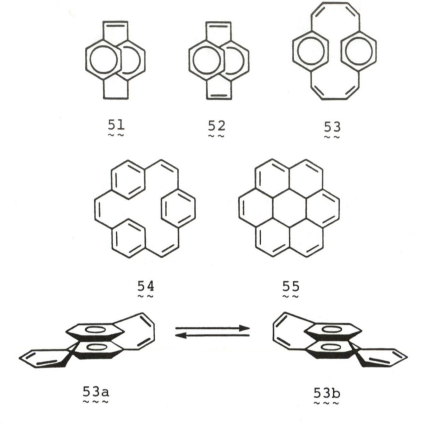

Sutherland, J. A. Reiss, H. A. Staab, F. Vogtle, and H. Hopf. Two
volumes were required to adequately review the subject in 1983.[46]

Carbanion Stereochemistry and Mechanism (1955–1972)

Origins

Ernest Grunwald told me after we had completed a series of over 50
full-length papers under the general title Electrophilic Substitution at
Saturated Carbon that had we entitled the series Carbanion Mechanisms
and Stereochemistry we would have increased our readership manifold.
However, we might have retrieved some readership through our monograph, *Fundamentals of Carbanion Chemistry*, in which the early results
were reviewed.[47]

The carbanion work represented our deepest incursion into the
study of organic reaction mechanisms. The research started with the

general question, "What are the stereochemical capabilities of carbanions?" We posed this question at the outset of 15 years of research for explicit reasons.

1. Carbanions were one of the three most important reaction intermediates in organic reactions, undoubtedly the most important intermediate in carbon—carbon bond-forming reactions in the biological world.

2. Carbanion stereochemistry had remained essentially unstudied up to 1955.

3. The application of stereochemical probes of the structures of carbonium ions and ion pairs had been rewarding in our discoveries of phenonium ions, phenonium tosylate ion-pair collapse (internal return), and anion-exchange reactions.

4. Physical—organic chemical research provided a wide opportunity for the exercise of what we enjoyed the most—the design of structures, the design of synthetic sequences to bring them to hand, the design of test systems that would yield scientific information, and the design of research problems that would provide co-workers with a wide familiarity with the thought patterns, techniques, and compound classes of organic chemistry.

When we started our work in 1954, the most serious probe of the stereochemical course of reactions involving carbanion intermediates had been made by C. L. Wilson, C. K. Ingold, S. K. Hsu, and their co-workers.[48,49] These investigators observed equal rates for hydrogen—deuterium exchange and racemization of optically active 56 in dioxane—D_2O—NaOD, and of 57 in H_2O—NaOH. They interpreted their results in terms of the intervention in these reactions of ambident anions that were planar, which were protonated from either face with equal probability, and that might be protonated on either oxygen or carbon. These early observations, made in 1936—1938, probably preserved the field from further investigation from 1938 to 1955 because planar carbanions are hardly susceptible to stereochemical scrutiny.

56   57

Reactions That Generate Carbanions

In my first study[50] with J. Allinger and A. Langemann, we carried out reverse condensation reactions of five different systems (58 is a prototype) in a variety of solvents catalyzed by various bases to produce 2-phenylbutane (59) and ketone (60) (aldehyde, acid, or amide in companion systems). The reactions were found to occur with as high as 93% net retention of configuration and up to 50% net inversion of configuration, depending mainly on the solvent. The highest retention was associated with 2-butanone (60) as the leaving group and t-C_4H_9OH–t-C_4H_9OK as the medium. The highest inversion involved the same leaving group, but with $HOCH_2CH_2OH$–$KOCH_2CH_2OH$ as the medium. We interpreted the results in terms of asymmetric ion-pairing and solvation of flat, or nearly planar carbanion intermediates. Later, when dimethyl sulfoxide-t-C_4H_9OK was used as medium, the 2-phenylbutane (59) produced was totally racemic (J. L. Mateos and K. R. Kopecky).[51]

$$C_6H_5 \overset{\overset{\displaystyle CH_3}{|}}{\underset{\underset{\displaystyle C_2H_5}{|}}{C}} \overset{\overset{\displaystyle O^- K^+}{|}}{\underset{\underset{\displaystyle CH_3}{|}}{C}} C_2H_5 \; + \; B\text{—}H \longrightarrow$$

$$\underset{\displaystyle 58}{}$$

$$C_6H_5 \overset{\overset{\displaystyle CH_3}{|}}{\underset{\underset{\displaystyle C_2H_5}{|}}{\overset{*}{C}}} \text{—H} \; + \; O{=}\overset{\overset{\displaystyle CH_3}{}}{\underset{\underset{\displaystyle CH_3}{|}}{C}}\text{—}C_2H_5 \; + \; \bar{B}\overset{+}{K}$$

$$\underset{\displaystyle 59}{} \qquad\qquad \underset{\displaystyle 60}{}$$

Systems were devised in which nitrogen (J. S. Bradshaw)[52] and oxygen leaving groups (C. A. Kingsbury)[53] also produced the 2-phenylbutyl anion, which in proton-donating media also gave 2-phenylbutane (59). Reactions 1–3 provide examples in which R* is the asymmetric center and represents the 2-phenylbutyl group. In reactions 1 and 2 with t-C_4H_9OH–t-C_4H_9OK as medium and KIO_3 or I_2 as oxidant (in reaction 2), the product was produced with approximately 80% net retention of configuration. In H_2O–KOH as medium, reaction 1 went with 5% net inversion. Reaction 1 carried out in H_2O–KOH with either KIO_3 or Br_2 as oxidant gave 59 with 33% net inversion. Reaction

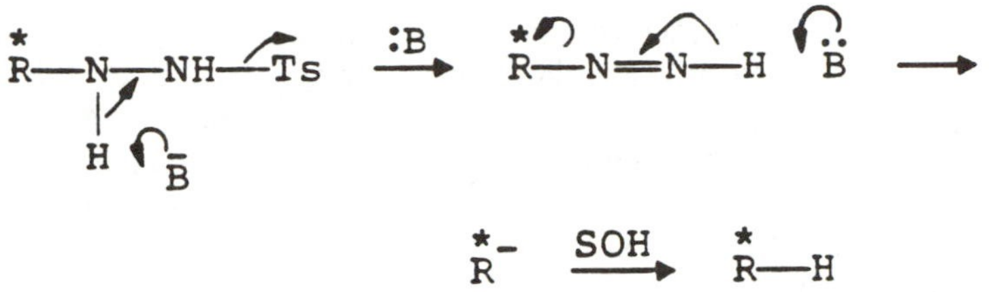

$$\overset{*}{R}-\underset{\underset{H}{|}}{N}-NH-Ts \quad \overset{:B}{\longrightarrow} \quad \overset{*}{R}-N=N-H \quad B \quad \longrightarrow$$

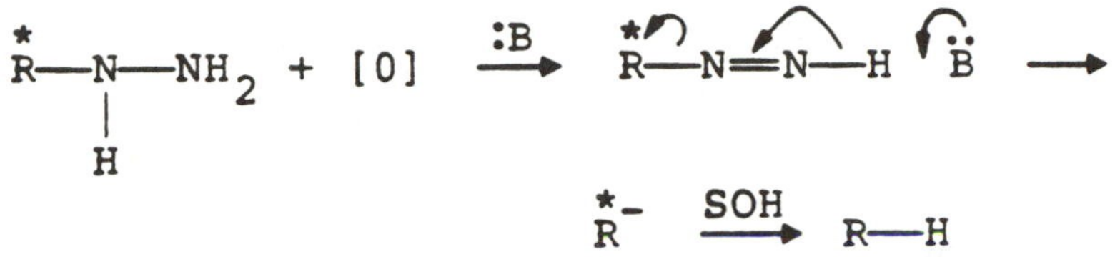

$$\overset{*}{R}^{-} \quad \overset{SOH}{\longrightarrow} \quad \overset{*}{R}-H \qquad\qquad (1)$$

$$\overset{*}{R}-\underset{\underset{H}{|}}{N}-NH_2 \; + \; [0] \quad \overset{:B}{\longrightarrow} \quad \overset{*}{R}-N=N-H \quad B \quad \longrightarrow$$

$$\overset{*}{R}^{-} \quad \overset{SOH}{\longrightarrow} \quad R-H \qquad\qquad (2)$$

$$\overset{*}{R}-O-CH-C_6H_5 \quad \longrightarrow \quad \overset{*}{R}^{-} \quad \overset{C_6H_5NHCH_3}{\longrightarrow} \quad R-H \quad (3)$$

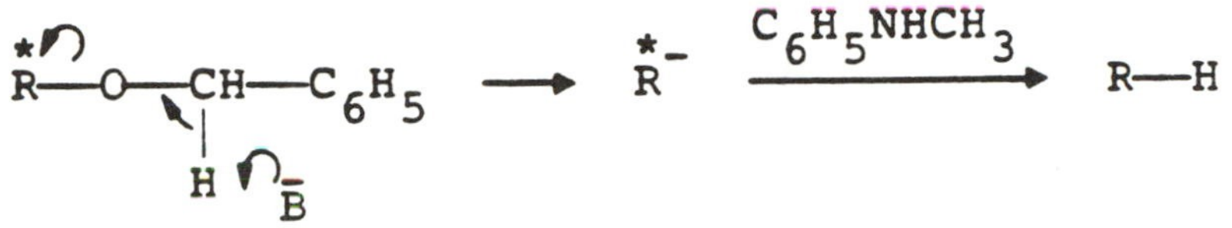

3 could be driven only with a very strong base in a high-boiling solvent $(C_6H_5(CH_3)NH-C_6H_5(CH_3)NK)$ under conditions that partially racemized the 2-phenylbutane formed. The reaction went with >29% retention.

In our first studies carried out with hydrogen or deuterium leaving groups in 1959[54] and 1961,[55], C. A. Kingsbury, B. Rickborn, and P. Haberfield and I carried out base-catalyzed isotopic exchange in the two systems 59h–59d and 61h–61d. In each reaction, the solvent carried the isotopic label opposite to that of the substrate. The results were similar to those obtained when the carbanions were generated with carbon, nitrogen, and oxygen leaving groups. For example, isotopic exchange of 59 and of 61 in t-$C_4H_9OH(D)$–t-C_4H_9OK went with approximately 95% net retention, in $(CH_3)_2SO$–t-C_4H_9OK with 100% racemization, and in $(D)HO(CH_2)_2O(CH_2)_2OH(D)$–$(D)HO(CH_2)_2O(CH_2)_2OK$, with 50–60% net inversion.

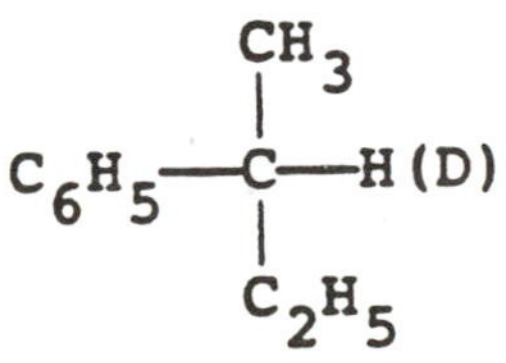

$$C_6H_5-\underset{\underset{C_2H_5}{|}}{\overset{\overset{CH_3}{|}}{C}}-H(D)$$

$$C_6H_5-\underset{\underset{OCH_3}{|}}{\overset{\overset{CH_3}{|}}{C}}-H(D)$$

59–h (59–d) 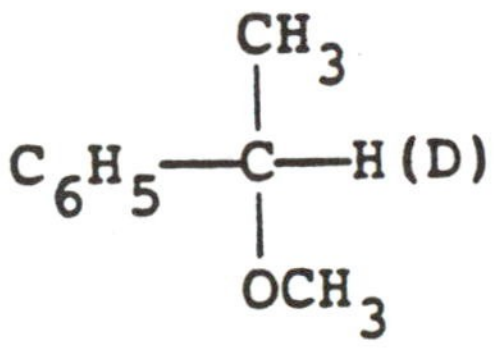   61–h (61–d)

The mechanisms suggested to accommodate these results involved planar carbanionic intermediates in either asymmetric or symmetric environments, depending on the medium. Formulas 62–64 represent the envisioned intermediates (L is a generalized leaving group). Intermediate 62 is an asymmetric ion pair of very short lifetime, formed in a nondissociating solvent such as t-C_4H_9OH. Proton donation to the carbanion on the face occupied by K^+ and L gives retention of configuration. This pathway is favored because the K^+OR^- formed by such a reaction is nondissociated. Proton capture from the opposite face to give inversion would have produced a higher energy product-separated ion pair $[RO^- \cdot abcCH \cdot K^+]$ in a nonionizing medium.

$$\underset{62}{\text{ROH}\cdots\overset{\overset{\text{HOR}}{|}}{\underset{\underset{\text{L}}{|}}{\underset{a}{\overset{b \quad c}{C}}}\cdots K^+}} \qquad \underset{63}{\text{ROH}\cdots\underset{\underset{a}{|}}{\overset{b \quad c}{C}}\cdots\text{L}} \qquad \underset{64}{\text{R}\!-\!\text{OH}\cdots\underset{\underset{a}{|}}{\overset{b \quad c}{C}}\cdots\text{HOR}}$$

Intermediate 63 is a short-lived carbanion, asymmetrically solvated by the leaving group on one face and by solvent on the other. It is formed in polar dissociating solvents such as $HOCH_2CH_2OH$ or H_2O, in which the basic catalyst is a dissociated anion. The species (63) collapses to give inverted product because of the shielding effect of L, which is a neutral molecule.

Symmetrically solvated intermediate 64 is formed whenever the polarity of the medium and the pK_a relationships between the medium and the product are such as to extend the lifetime of the carbanion to the point where collapse to the covalent state is slow compared to the symmetrization processes.[56]

Chiral Carbon Acids

A kinetic approach to the study of the stereochemical course of carbanion-producing reactions involved determination and comparison of the relative rates of base-catalyzed hydrogen–deuterium isotopic exchange (k_e) and racemization (k_α) of carbon acids that are asymmetric at the acidic site. Valid mechanistic conclusions can be drawn from k_e/k_α values only if the studies are carried out in the presence of isotopic pools of acidity comparable to that of the conjugate acid of the base used to catalyze the reaction. Four limiting ratios are envisioned.

1. If isotopic exchange occurs with complete *retention* of configuration, $k_e/k_\alpha = \infty$.

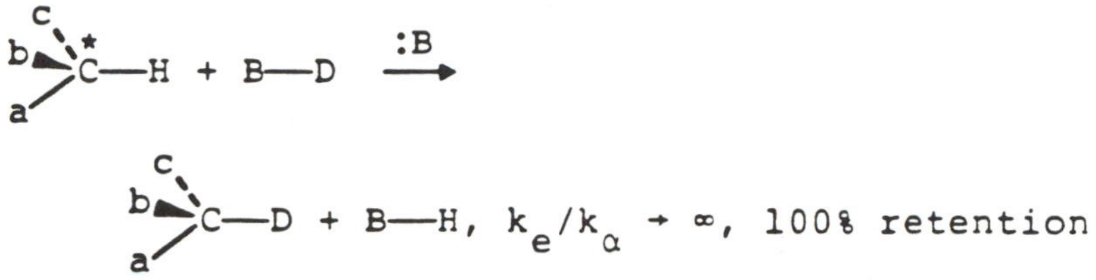

2. If exchange occurs with complete *racemization*, $k_e/k_\alpha = 1$.

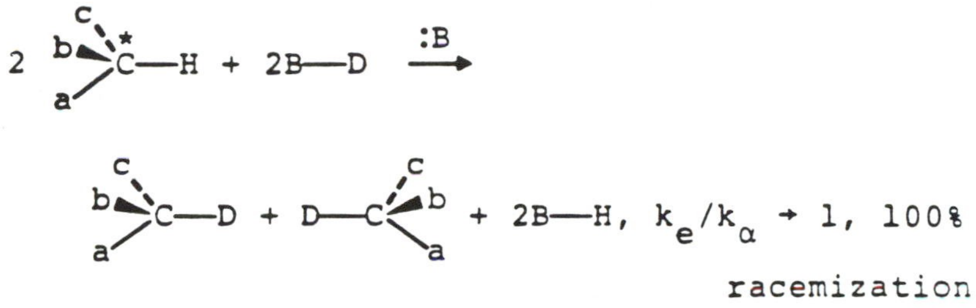

3. If exchange goes with complete *inversion* of configuration, each carbanion formed produces material of a sign opposite to that of the starting material. After half of the material is exchanged, the rotation becomes zero. Thus complete inversion provides $k_e/k_\alpha = 0.5$.

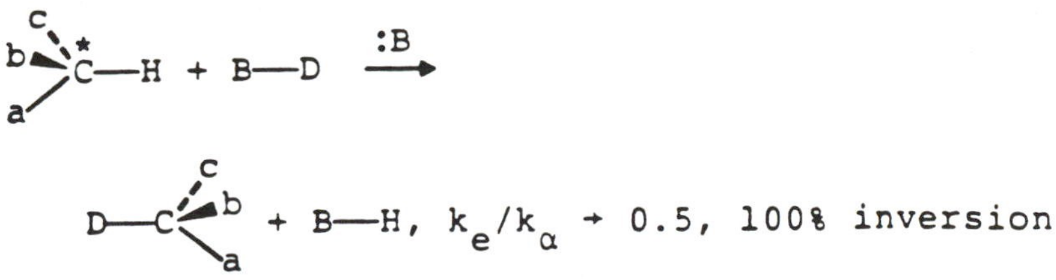

4. If racemization occurs without isotopic exchange, $k_e/k_\alpha = 0$. Such a process involves proton or deuteron transfer from one face to the opposite face of a carbanion without isotopic exchange with material in the pool. We referred to such a result as *isoracemization* that goes by an *isoinversion* mechanism.

$$\overset{c}{\underset{a}{\overset{b}{\diagdown}}}\overset{*}{C}-H \;+\; B-D \;\xrightarrow{:B}\;$$

$$H-\overset{c}{\underset{a}{\overset{b}{C}}} \;+\; B-H, \quad k_e/k_\alpha \to 0, \quad 100\% \text{ isoinversion}$$

By 1964, L. Gosser and I[57,58] had realized all four of the envisioned results experimentally in fluorene systems 65 and 66. The mechanisms used to interpret these results were formulated. In all cases, the fluorenyl anion was likely to be planar, but as with planar carbonium ions, it was capable of either entering into asymmetric ion pairs, or of being asymmetrically solvated.

65   66

Substitution with retention of configuration (e.g., $k_e/k_\alpha = 50$) was observed with 65 in nonpolar solvents such as t-C_4H_9OH with NH_3 as the base. A similar result was observed in C_6H_6:C_6H_5OH–C_6H_5OK (90:10) ($k_e/k_\alpha = 18$). In the mechanisms envisioned for the reactions in equations 4 and 5, rotations of the cations of the contact ion-pair intermediates and collapses to the covalent states are faster processes than either ion-pair dissociations or anion rotations within the ion pairs. Although the intermediate carbanions are probably hydrogen-bonded on the face remote from the leaving group, the pK_a relationships and the formation of *product-separated* ion pairs inhibit collapses to give inverted product.

When the more polar solvents CH_3OH or $(CH_3)_2SO$ were substituted for t-C_4H_9OH in the ammonia-catalyzed reaction of equation 1, $k_e/k_\alpha = 1$. Similarly, when $(CH_3)_4NOC_6H_5$ was substituted for KOC_6H_5 in the same medium as that of equation 5, $k_e/k_\alpha = 1$. In the first case, we concluded that the ammonium carbanionide ion pair initially formed dissociated in the polar solvents to produce symmetrically solvated carbanions faster than ion-pair reorganization and collapse had occurred. In the second case, the absence of a ligating cation in the ion pair destroyed the retention–reaction pathway, and thus allowed the competing ion-pair dissociating pathway to operate. Equation 6 generalizes the racemization mechanism (W. T. Ford).[59]

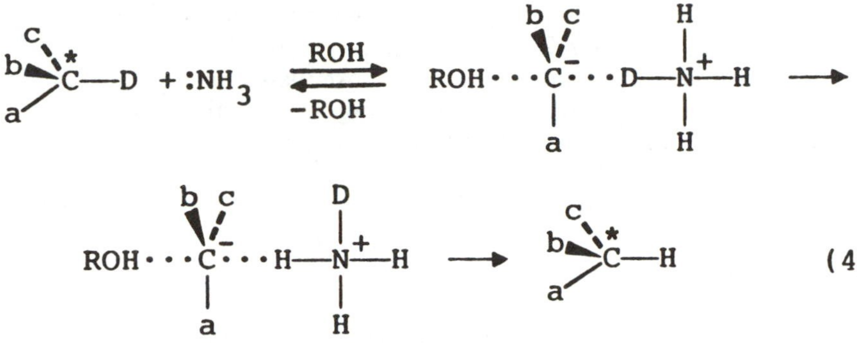

$$\underset{a}{\overset{c}{\underset{b}{\diagup}}}\!\!\overset{*}{C}\!-\!D \;+\; :NH_3 \;\underset{-ROH}{\overset{ROH}{\rightleftarrows}}\; ROH\cdots\underset{a}{\overset{b\quad c}{C}}\cdots D\!-\!\overset{H}{\underset{H}{\overset{+}{N}}}\!-\!H \;\longrightarrow$$

$$ROH\cdots\underset{a}{\overset{b\quad c}{C}}\cdots H\!-\!\overset{D}{\underset{H}{\overset{+}{N}}}\!-\!H \;\longrightarrow\; \underset{a}{\overset{c}{\underset{b}{\diagup}}}\!\!\overset{*}{C}\!-\!H \qquad (4)$$

$$\underset{a}{\overset{c}{\underset{b}{\diagup}}}\!\!\overset{*}{C}\!-\!D \;+\; K^+ \begin{matrix}\bar{O}C_6H_5\\ \vdots\\ \vdots\\ HOC_6H_5\end{matrix} \;\underset{-ROH}{\overset{ROH}{\rightleftarrows}}\; ROH\cdots\underset{a}{\overset{b\ c}{C}}\cdots DOC_6H_5 \;\begin{matrix}\vdots\\ K^+\\ \vdots\\ HOC_6H_5\end{matrix} \;\longrightarrow$$

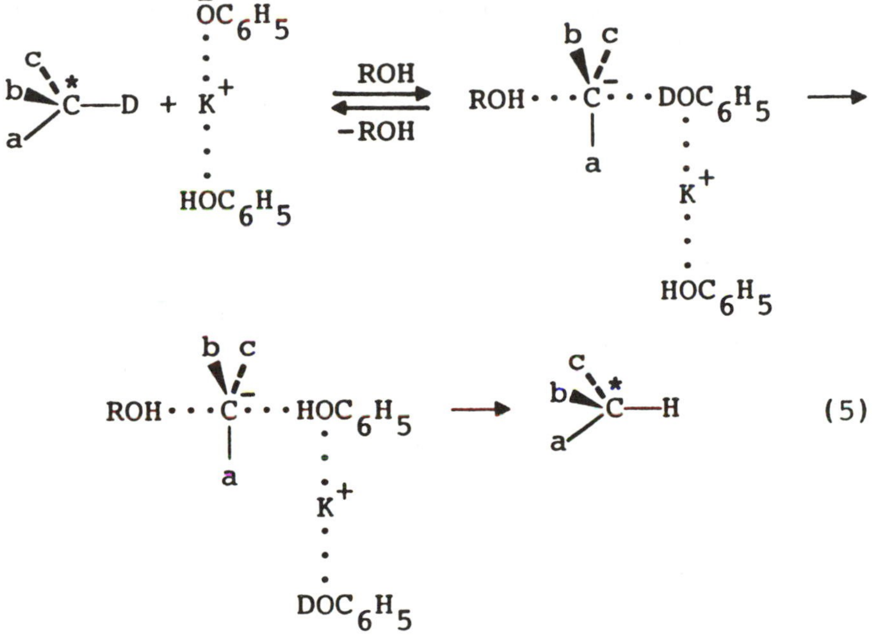

$$ROH\cdots\underset{a}{\overset{b\ c}{C}}\cdots HOC_6H_5 \;\begin{matrix}\vdots\\ K^+\\ \vdots\\ DOC_6H_5\end{matrix} \;\longrightarrow\; \underset{a}{\overset{c}{\underset{b}{\diagup}}}\!\!\overset{*}{C}\!-\!H \qquad (5)$$

$$\underset{a}{\overset{c}{\underset{b}{\diagup}}}\!\!\overset{*}{C}\!-\!D \;+\; BH \;+\; :B \;\rightleftharpoons\; BH\cdots\underset{a}{\overset{b\ c}{C}}\cdots DB \;\underset{-BD}{\overset{+BH}{\longrightarrow}}$$

$$BH\cdots\underset{a}{\overset{b\ c}{C}}\cdots HB \;\longrightarrow\; \underset{a}{\overset{c}{\underset{b}{\diagup}}}\!C\!-\!H \;+\; H\!-\!\underset{a}{\overset{c}{\underset{b}{\diagdown}}}C \qquad (6)$$

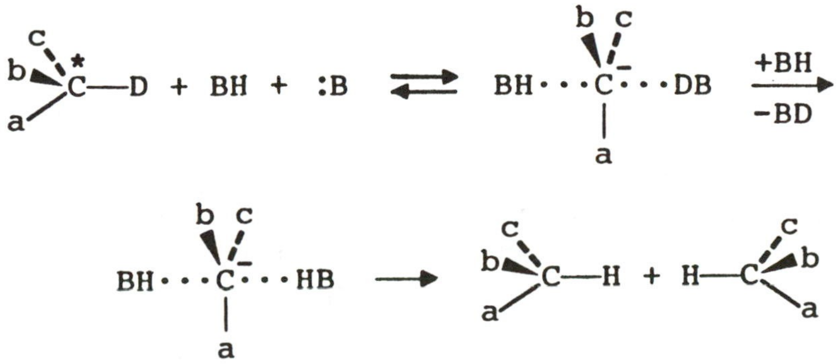

When **65** was treated with $CH_3OH–KOCH_3$ or with $CH_3OH–(n-C_3H_7)_3N$, k_e/k_α values of 0.78 and 0.65, respectively, were observed.[60] The predominant inversion pathway is formulated in equation 7. In this mechanism, dissociated methoxide anion is the active base, and ionization is aided by hydrogen bonding on the side of the incipient carbanion remote from the leaving proton to give an unsymmetrically solvated carbanion. Deuteron capture on one side regenerates starting material; proton capture on the opposite side provides inverted material, as indicated in equation 7.

Isoracemization with a k_e/k_α ratio of 0.1 was observed by W. T. Ford et al.[60] when **66** was treated with $(n-C_3H_7)_3N$ in $(CH_2)_4O-t-C_4H_9OH$ in the presence of $(n-C_3H_7)_3NHI$ that acted as a possible proton reservoir. The gross mechanism envisioned for this type of result is formulated in equation 8.

$$
\begin{array}{c}
\overset{c}{\underset{a}{\overset{b}{\diagup}}}\!\!\overset{*}{C}\!\!-\!\!D + CH_3OH + CH_3\bar{O} \rightleftharpoons
\end{array}
$$

$$
CH_3OH\cdots\overset{b\;c}{\underset{a}{\overset{\diagup}{C}}}{}^{\!\!-}\cdots DOCH_3 \quad\xrightarrow[{-CH_3O^-}]{{-CH_3OD}}\quad H\!-\!\overset{*}{C}\overset{c}{\underset{a}{\diagdown b}} \qquad (7)
$$

$$
\overset{c}{\underset{a}{\overset{b}{\diagup}}}\!\!\overset{*}{C}\!\!-\!\!D + :NPr_3 \longrightarrow \overset{b\;c}{\underset{a}{\overset{\diagup}{C}}}{}^{\!\!-}\cdots DNPr_3^{+} \longrightarrow
$$

$$
Pr_3\overset{+}{N}D\cdots\overset{b\;c}{\underset{a}{\overset{\diagup}{C}}}{}^{\!\!-} \longrightarrow D\!-\!\overset{*}{C}\overset{c}{\underset{a}{\diagdown b}} + :NPr_3 \qquad (8)
$$

Conducted Tour Mechanism for Isoinversion

Although **67** also exhibited isotopic exchange with retention, inversion, and racemization under the appropriate conditions, isoracemization was more favored than with **66**. For example, $k_e/k_\alpha = 0.19$ when **67** was treated with $(n-C_3H_7)_3N$ in $(CH_2)_4O-t-C_4H_9OH$ (0.10 N) in $(n-C_3H_7)_3NHI$. This result led to the proposal of a *conducted tour mechanism* for isoinversion. In this mechanism, drawn first in 1964,[61] the proton is abstracted by the amine base to generate a hydrogen-bonded ion pair whose negative charge delocalizes into a π system, along which the hydrogen-bonded ammonium ion can travel without serious charge separation. When negative charge distributes on nitrogen or on oxygen atoms that already contain other unshared electron pairs (as in 65–67), the proton can move from one face to the opposite face of the planar carbanion without the hydrogen bond being fully broken. If this process and that of ion-pair collapse is fast compared to competing reac-

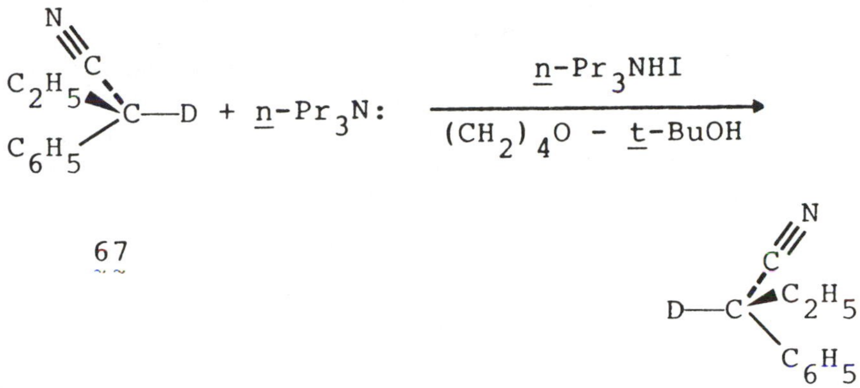

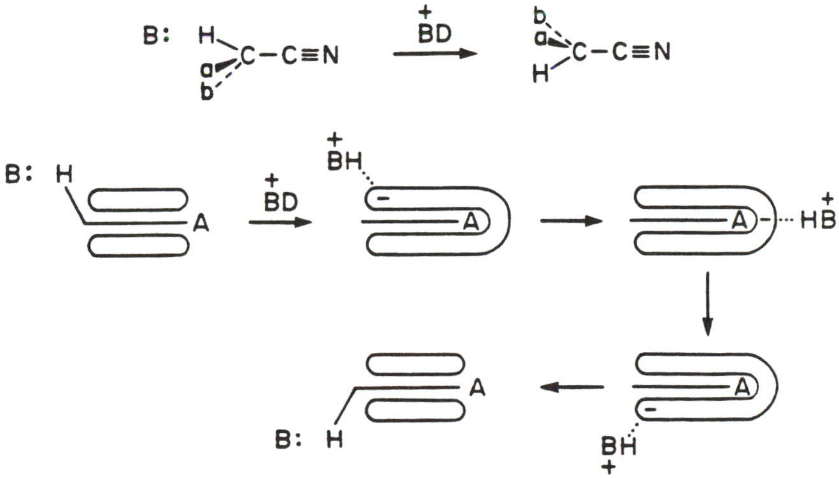

Isoinversion

Scheme I. Representation of a conducted tour mechanism for isoinversion.

tions, isoracemization occurs. In effect, the carbanion of an ammonium carbanionide ion pair rotates with respect to the ammonium group to produce isoinverted product. Scheme I is a representation of a *conducted tour mechanism for isoinversion*.[61]

Conducted Tour Mechanism for 1,3- and 1,5-Proton Shifts

By 1959, we had reported[54] that 59 and 61 underwent base-catalyzed isotopic exchange with high retention of configuration in t-C_4H_9OH–t-C_4H_9OK. By 1960, 68 had been found to give the same result.[62] All three systems exhibited exceptionally low hydrogen–deuterium isotope effects for the racemization and isotopic exchange reactions, a result well-established by 1961 by D. A. Scott, W. D. Nielson, and me.[63] The values ranged from 0.3 to 3.0; these low isotope effects coupled with the stereochemical results indicated that intermediate ion pairs collapsed to the covalent state faster than other detectable events occurred. In reaction 9, when $k_{-1} > k_2$ or k_3, then $k_e^{obs} = K(k_2 + k_3)$ and $k_\alpha^{obs} = Kk_3$, in which K is the equilibrium constant defined by $K = k_1/k_{-1}$. Thus, the observed isotope effects involved much more than k_1^H/k_1^D as had been previously assumed. These isotope effects were the products of equilibrium constants ($K = k_1/k_{-1}$) and rate constants that involved the

$$C_6H_5 \overset{\displaystyle O}{\underset{\displaystyle O}{\overset{\|}{\underset{\|}{S}}}} \overset{\displaystyle CH_3}{\underset{\displaystyle H(D)}{\overset{|}{\underset{|}{C}}}} C_6H_{13}\text{-}\underline{n}$$

$$\underset{\sim\sim}{68}$$

$$\overset{*}{R}\text{—}H + \bar{B}\overset{+}{K} \quad \overset{k_1}{\underset{k_{-1}}{\rightleftarrows}}$$
$$\vdots$$
$$DB$$

$$\overset{*}{R}\bar{}\cdots HB\cdots \overset{+}{K} \quad \overset{k_2}{\longrightarrow} \quad \overset{*}{R}\bar{}\cdots DB\cdots \overset{+}{K} \quad \overset{BH}{\vdots} \quad \longrightarrow \quad \overset{*}{R}\text{—}D \text{ (retention)} \quad (9)$$
$$\vdots$$
$$DB$$

$$k_3\Big\downarrow DB$$
$$BD\cdots \overset{*}{R}\bar{}\cdots DB \quad \longrightarrow \quad R\text{—}D \text{ (racemization)}$$

exchange of one hydrogen-bonded carbanion for another. A clear implication of $k_{-1} > k_2$ or k_3 was that a system prone to base-catalyzed rearrangement such as 69→70 should occur intramolecularly, without or with only partial isotopic exchange with the medium.

My experiments with R. T. Uyeda[64] demonstrated that 69 rearranged to 70 with a minimum of 54% of intramolecularity. This result was interpreted in terms of *anionic* reorganization within a potassium carbanionide ion-pair intermediate occurring competitively with *cationic* reorganization. This first report[64] of a base-catalyzed, intramolecular proton transfer was published in 1962. It was followed in 1963 by reports by G. Bergson et al.,[65] A. Schriesheim et al.,[66] R. B. Bates et al.,[67] and W. von E. Doering et al.,[68] all of whom also observed intramolecular base-catalyzed allylic rearrangements by 1,3-proton transfers. The observations of Bergson et al.[65] were particularly elegant, and made use, for the first time, of the relatively high acidity and interesting symmetry properties of the carbon acids of the indene systems, 71–73. Interestingly, enzyme-catalyzed intramolecular proton transfers were observed as early as 1960.[69–71]

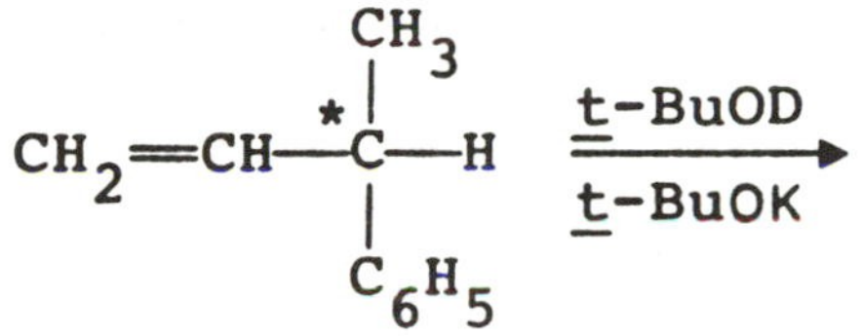

69

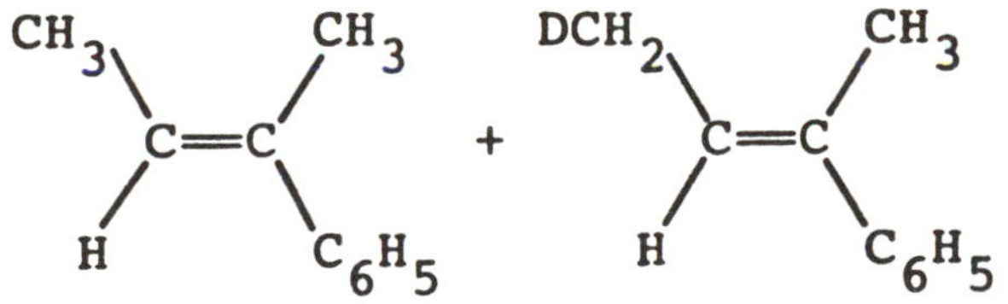

70   70-d

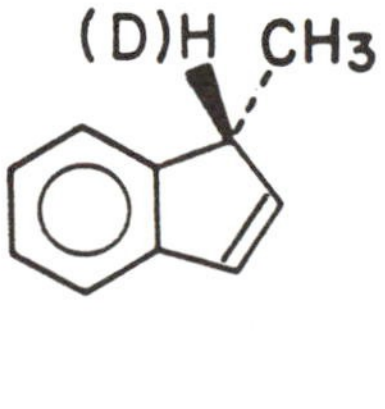

71 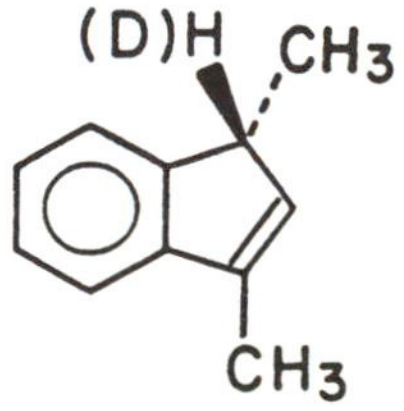   72 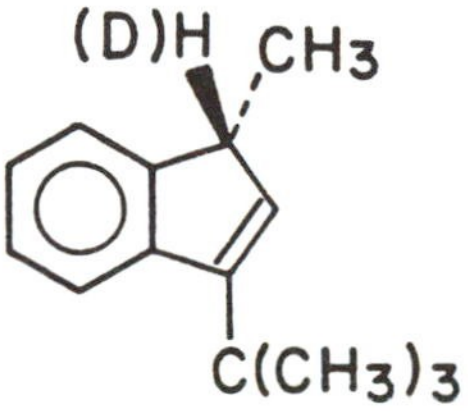   73

The fact that $k_{-1} > k_2$ or k_3 (reaction 9) in many systems compli-
cates the application of hydrogen isotopic exchange rates to extensions
of pK_a scales for very weak carbon acids and to Brønsted relationships.
The consequences of intervention of *internal return* in ionization to produce carbanions in ion pairs proved to have many parallels with the
consequences in ionization to produce carbonium ions in ion pairs.[72]

H. P. Fischer, D. A. Scott, F. Willey, and I demonstrated[73] in 1964
that **74** underwent intramolecular base-catalyzed 1,5-proton transfers to
give **75** in a variety of media. For example, in CH_3OD-CH_3ONa, the
reaction was 47% intramolecular; in $(CD_3)_2SO-CH_3OD-CH_3OK$ (10:90
ratio of $(CD_3)_2SO$ to CH_3OD) it was 40% intramolecular; in
$(C_2H_5)_3COD-(n-C_3H_7)_3N-(n-C_3H_7)_3NDI$, it was 98% intramolecular.
These reactions were interpreted[74] as occurring by *conducted* tour
mechanisms along the same face of the carbanion from which the proton was originally abstracted (*see* Scheme II).

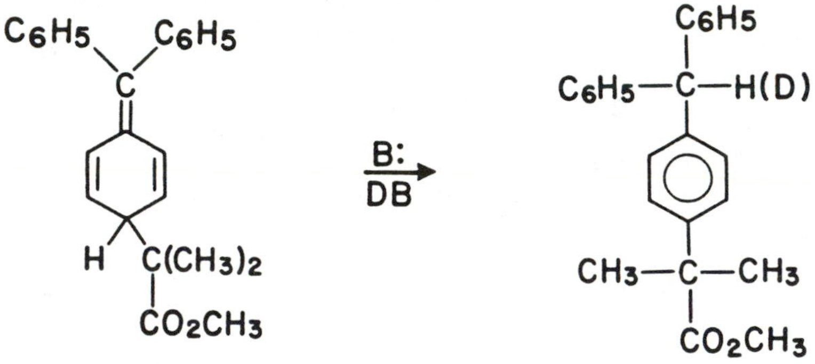

$$ \underset{74}{\sim\sim} \qquad\qquad\qquad\qquad \underset{75}{\sim\sim} $$

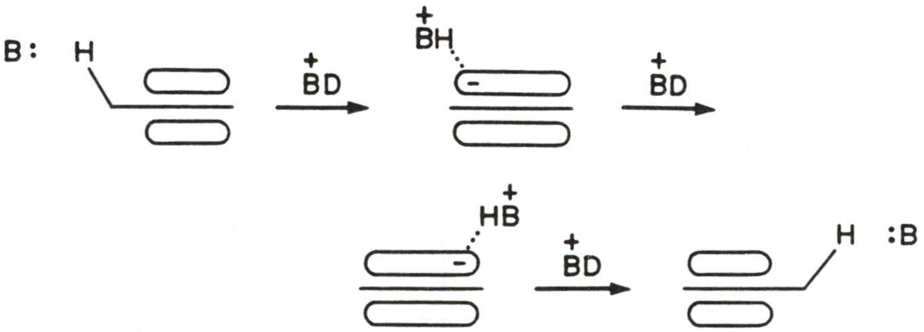

*Scheme II. Conducted tour mechanism for base-catalyzed 1,3- and
1,5-proton transfers.*

1,3-Asymmetric Induction by Conducted Tour Mechanisms

Bergson's 1963 studies[65] strongly suggested and supported the idea that, in appropriate systems, high 1,3-asymmetric induction can accompany base-catalyzed 1,3-rearrangements. Bergson's system **73** and its rearrangement product **76** had ideal structural features for studies of carbanion ion-pair behavior as a function of cation, solvent, base, and isotope effects. In a series of studies with R. T. Uyeda, J. Almy, D. C. Garwood, D. H. Hoffman, and K. C. Chu,[75-78] we determined the configuration and examined the rates of racemization, isotopic exchange, and rearrangement of **73** and **76** in a variety of solvents and with a variety of bases. The conducted tour mechanism for 1,3-asymmetric induction was established. We also observed that the presence of crown ether **77** in the medium increased the rates of some processes and depressed the rates of others. When pentamethylguanidine was used as base, an example of isoracemization without a conducted tour pathway was uncovered. This result was interpreted as being due to carbanion rotation within an ion pair whose cation's positive charge was very highly delocalized.[78]

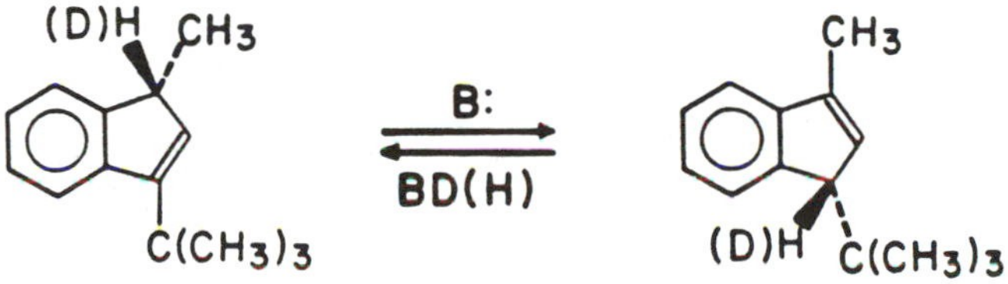

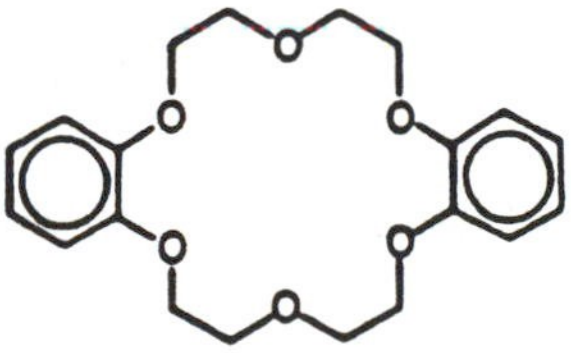

Base-Catalyzed Azaallyic Rearrangements

In C. K. Ingold's classic book, *Structure and Mechanism in Organic Chemistry*,[79] he stated, as a result of extensive kinetic studies of the base-catalyzed rearrangement of **78** to **79** (and other systems), "it proves the one-stage bimolecular mechanism. . . ." This conclusion conflicted with our findings that the base-catalyzed rearrangements of **69** and **74** involved carbanion intermediates. Because of this conflict, and the importance of the azaallylic rearrangement to the transamination reactions of biochemistry, R. D. Guthrie and I[80] (1966) studied the rearrangement of **80** to **81**. The results clearly established the intervention of carbanion ion pairs, of internal return, and intramolecularity in the reactions.

$$\text{p-C}_6\text{H}_5\text{C}_6\text{H}_4\text{---}\overset{\overset{\displaystyle \text{C}_6\text{H}_5}{|}}{\underset{\underset{\displaystyle \text{H}}{|}}{\overset{*}{\text{C}}}}\text{---N}\text{==}\overset{\overset{\displaystyle \text{C}_6\text{H}_5}{|}}{\text{C}}\text{---}\text{C}_6\text{H}_5 \quad \xrightarrow{\text{Base}}$$

78

$$\text{p-C}_6\text{H}_5\text{C}_6\text{H}_4\text{---}\overset{\overset{\displaystyle \text{C}_6\text{H}_5}{|}}{\text{C}}\text{==N}\text{---}\overset{\overset{\displaystyle \text{C}_6\text{H}_5}{|}}{\underset{\underset{\displaystyle \text{H}}{|}}{\text{C}}}\text{---}\text{C}_6\text{H}_5$$

79

Warren T. Ford, T. J. Leitereg, Donald J. Cram, S. M. Wong, S. Wong, R. E. Singler, T. Whitney, J. Almy, and J. N. Roitman at the Wongs' wedding. After the wedding, the entire party went to a Chinese restaurant.

$$
\begin{array}{c}
\overset{\displaystyle CH_3}{\underset{\displaystyle H}{\overset{|}{\underset{|}{C_6H_5\!-\!C}}}}\!-\!N\!\!=\!\!\overset{\displaystyle C_6H_4Cl\text{-}\underline{p}}{\overset{|}{C}}\!-\!C_6H_4Cl\text{-}\underline{p} \quad \overset{:B}{\rightleftarrows}
\end{array}
$$

$$\underset{\sim\sim}{80}$$

$$
\begin{array}{c}
\overset{\displaystyle CH_3}{\overset{|}{C_6H_5\!-\!C}}\!\!=\!\!N\!-\!\overset{\displaystyle C_6H_4Cl\text{-}\underline{p}}{\underset{\displaystyle H}{\overset{|}{\underset{|}{C}}}}\!-\!C_6H_4Cl\text{-}\underline{p}
\end{array}
$$

$$\underset{\sim\sim}{81}$$

R. D. Guthrie, D. A. Jaeger, W. Meister, and I[81] extended the study to the conversion of 82 to 83. The conversion of 84 to 85 that I performed with D. A. Jaeger[82] indicated that the reactions occur with 97–100% stereospecificity and with lower, but substantial, intramolecularity in $t\text{-}C_4H_9OH\!-\!t\text{-}C_4H_9OK$. In a competing reaction, 82 and 84 undergo isotopic exchange with the medium with high retention of configuration. Addition of crown ether 77 to the medium completely destroys the stereospecificity of the isotopic exchange reaction. Again, ion-pair reorganization reactions (anion and cation rotations within the ion pair) explain the results. The very high asymmetric induction observed in the 1,3-rearrangements is dependent on the absence of ion pairs of configurations or conformations other than A and B as intermediates. The high steric requirements of the C_6H_5 and $(CH_3)_3C$ groups account for the stability of A and B, relative to that of competing conformations.

In a system that more closely models enzyme-moderated transaminations, M. Broadhurst, D. A. Jaeger, and I[83] examined the isomerization of 86 to 87. The rearrangements occurred with 12–29% stereospecificity and 37% intramolecularity. The isotopic exchange of 86 went with isoracemization, presumably by a conducted tour mechanism around the pyridyl nitrogen ($k_e/k_\alpha = 0.25$), whereas that of 87 occurred with rather high retention of configuration ($k_e/k_\alpha = 7$).

In interpreting the results of exploratory research, different scientists exhibit differing degrees of timidity. No one likes to be wrong, and everyone likes to discover generally useful new principles that are illustrated for the first time by their work. My former colleague, Saul Winstein, hated to be wrong. He disliked even more for others to be

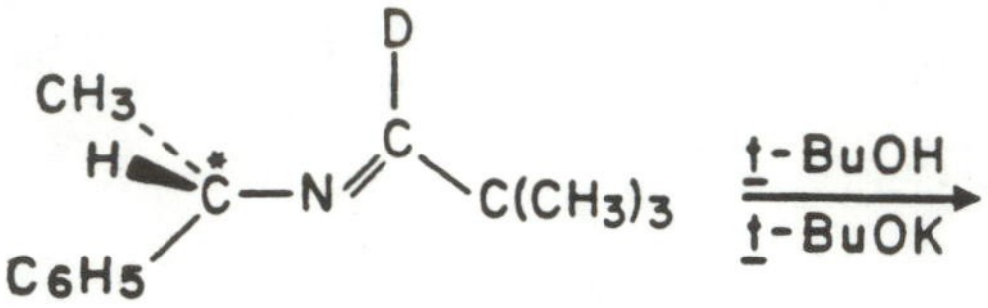

$$\underset{82}{}$$

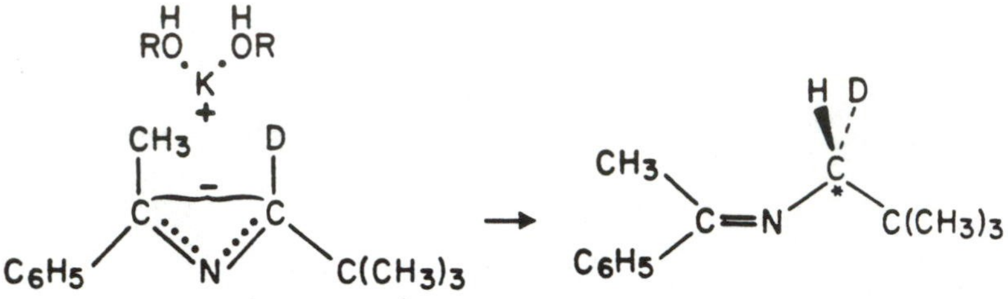

A (K$^+$ and ligands on top face)

$$\underset{83}{}$$

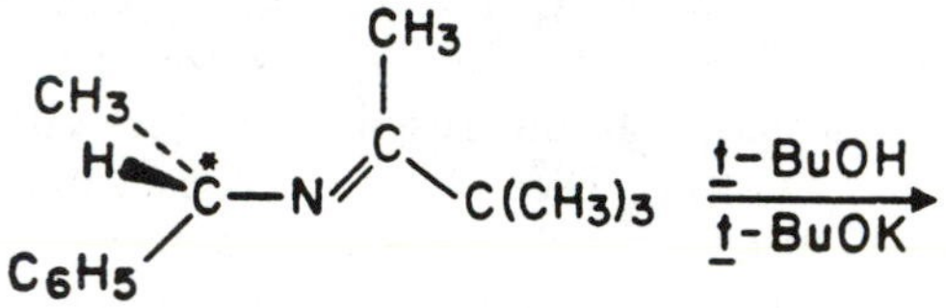

$$\underset{84}{}$$

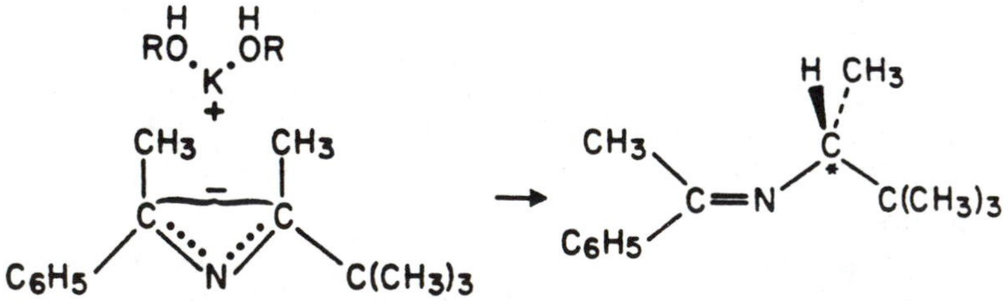

B (K$^+$ and ligands on top face)

$$\underset{85}{}$$

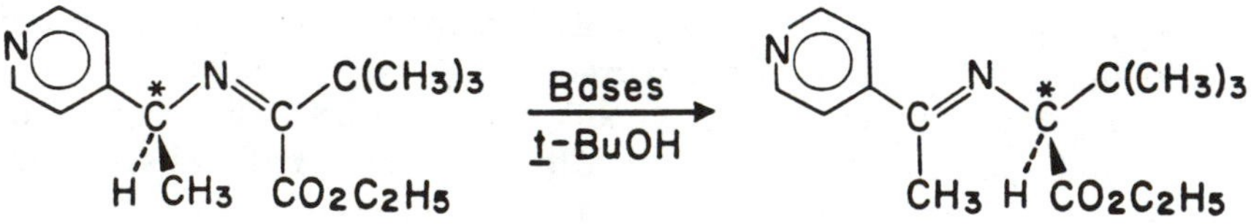

$$\underset{86}{} \qquad\qquad \underset{87}{}$$

wrong. He always enthusiastically cross-examined seminar speakers at UCLA to see if their interpretations were unique.

When I first proposed the conducted tour mechanism for proton transfer, I gave a 3-hour seminar on the idea at UCLA. It quickly expanded to four 3-hour seminars in which every detail was "scrutinized" (Winstein's favorite word). The resulting inquisition, although painful, was very profitable. To the extent my pride was hurt, my interpretations were improved. My resourcefulness was at its peak during these battles, and the resulting stimulation (sometimes accompanied by humiliation) was invaluable. Anyone who wishes to grow over their entire lifetime needs criticism. My wife, Jane M. Cram, has been my most inspiring and unsparing critic since Saul Winstein's untimely death.

Asymmetric Carbanions and Their Environments

The studies of all of the systems discussed thus far involved carbanions that are intrinsically planar, and therefore nonchiral. Three reports of the unique behavior of 68, when subject to base-catalyzed isotopic exchange with solvent, appeared almost simultaneously. In the first[62] D. W. Nielsen, B. Rickborn, and I (1960) contrasted the stereochemical course of base-catalyzed isotopic exchange of 68, 59, 61, and 88 with solvents of the opposite isotopic type. Compound 68d underwent isotopic exchange with high retention of configuration in $t\text{-}C_4H_9OH\text{-}t\text{-}C_4H_9OK$, $HO(CH_2)_2OH\text{-}HO(CH_2)_2OK$, $CH_3OH\text{-}CH_3OK$, and $CH_3SOCH_3\text{-}CH_3OH\text{-}CH_3OK$. Compounds 59d and 61d gave isotopic exchange

$$
\begin{array}{cc}
\underset{\underset{68}{\sim\sim}}{
\overset{\displaystyle O}{\underset{\displaystyle O}{C_6H_5\!-\!\overset{\|}{\underset{\|}{S}}\!-\!\overset{\displaystyle CH_3}{\underset{\displaystyle C_6H_{13}\text{-}\underline{n}}{\overset{*}{C}}}\!\!-\!\!H\,(D)}}
&
\underset{\underset{59}{\sim\sim}}{C_6H_5\!-\!\overset{\displaystyle CH_3}{\underset{\displaystyle C_2H_5}{\overset{*}{C}}}\!\!-\!\!H\,(D)}
\\[3em]
\underset{\underset{61}{\sim\sim}}{C_6H_5\!-\!\overset{\displaystyle CH_3}{\underset{\displaystyle OCH_3}{C}}\!\!-\!\!H\,(D)}
&
\underset{\underset{88}{\sim\sim}}{C_6H_5\!-\!\overset{\displaystyle CH_3}{\underset{\displaystyle CONEt_2}{\overset{*}{C}}}\!\!-\!\!H}
\end{array}
$$

with high retention of configuration in $t\text{-}C_4H_9OH\text{-}t\text{-}C_4H_9OD$ because of asymmetric ion pairing, with moderate inversion of configuration in $O(CH_2CH_2OH)_2\text{-}HO(CH_2)_2O(CH_2)_2OK$ because of asymmetric solvation of non-ion-paired carbanion. Compound 61d gave total racemization in $(CH_3)_2SO\text{-}2$ M $t\text{-}C_4H_9OH\text{-}t\text{-}C_4H_9OK$ because of fast ion-pair dissociation in that polar medium.[54] Compound 88 gave complete racemization in $t\text{-}C_4H_9OD\text{-}t\text{-}C_4H_9OK$, because of the ambident character of the α-carbonyl carbanion.

We concluded that the carbanion from 68 was asymmetric either because of a "d-orbital-stabilized sp^3 hybrid with exchange occurring faster than inversion, or as a rehybridized, but still asymmetric, species in which the carbanion is 'doubly bonded' to sulfur by $p\text{-}d$ orbital overlap. In the latter case, formation and disposal of the carbanion might occur preferentially from the same side because of steric and(or) dipole–dipole interactions."[62]

In a second and independent study, E. J. Corey and E. T. Kaiser (1961)[84] reported that 68d underwent isotopic exchange in $EtOH\text{-}H_2O$ (2:1) with high retention of configuration because of carbanion formation (either sp^2 or sp^3) while in an asymmetric conformation, which is maintained throughout the isotopic exchange. In a third independent study, H. L. Goering et al.[85] reported high k_e/k_α values for 68 in methanol. As a result of these observations, D. A. Scott, W. D. Nielsen, and I[63] undertook a detailed kinetic study of k_e/k_α values of 68 in different solvents. The values varied from as high as 2000 to as low as 10 as media were changed in the following order: $t\text{-}C_4H_9OH\text{-}t\text{-}C_4H_9OK$ > $t\text{-}C_4H_9OD\text{-}t\text{-}C_4H_9OK$ >> $t\text{-}C_4H_9OH\text{-}(CH_3)_4NOH$ > $HOCH_2CH_2OH\text{-}HOCH_2CH_2OK$ > $DOCH_2CH_2OD\text{-}DOCH_2CH_2OK$ > $CH_3OH\text{-}CH_3OK$ ~ $(CH_3)_2SO\text{-}CH_3OH\text{-}CH_3OK$. Clearly asymmetric ion-pairing and solvation effects are imposed on α-sulfonylcarbanions that are, themselves, asymmetric. In another study with A. S. Wingrove, 89 and 90 were found to undergo base-catalyzed cleavage reactions to give 68 with high retention of configuration.[86,87]

In further experiments with R. D. Partos,[88] S. H. Pine,[89] R. D. Trepka,[90] and P. St. Janiak,[90] the effects of other d-orbital-stabilizing

89

90

groups on carbanion configuration were examined through k_e/k_α determinations in the same types of media used for **68**. The seven systems fall into two classes: those like sulfone **68** that give k_e/k_α values of 10 or higher irrespective of the medium used (**91–93**, class A) and those that give k_e/k_α values of 1–3, irrespective of the medium used (**94–97**, class B). The structural feature common to the class A compounds is that they all contain two unsubstituted oxygens bonded to the second-row element. The class B compounds all contain either one or three unsubstituted oxygens bonded to the second-row element.[88–90]

$$
\begin{array}{c}
\overset{\displaystyle CH_3}{\underset{\displaystyle}{|}} \quad \overset{\displaystyle O}{\underset{\displaystyle\|}{}} \quad \overset{\displaystyle CH_3}{\underset{\displaystyle}{|}} \\
C_6H_5 - N - S \overset{*}{-} C - H\,(D) \\
\overset{\displaystyle\|}{\underset{\displaystyle O}{}} \quad \overset{\displaystyle}{\underset{\displaystyle C_6H_{13}\text{-}\underline{n}}{}}
\end{array}
$$

91

$$
\begin{array}{c}
\overset{\displaystyle O}{\underset{\displaystyle\|}{}} \quad \overset{\displaystyle CH_3}{\underset{\displaystyle}{|}} \\
CH_3 - O - S \overset{*}{-} C - H \\
\overset{\displaystyle\|}{\underset{\displaystyle O}{}} \quad \overset{\displaystyle}{\underset{\displaystyle C_6H_{13}\text{-}\underline{n}}{}}
\end{array}
$$

92

$$
\begin{array}{c}
\overset{\displaystyle KO}{\underset{\displaystyle}{|}} \quad \overset{\displaystyle CH_3}{\underset{\displaystyle}{|}} \\
C_6H_5 - P \overset{*}{-} C - H \\
\overset{\displaystyle\|}{\underset{\displaystyle O}{}} \quad \overset{\displaystyle}{\underset{\displaystyle C_6H_{13}\text{-}\underline{n}}{}}
\end{array}
$$

93

$$
\begin{array}{c}
\overset{\displaystyle O}{\underset{\displaystyle\|}{}} \quad \overset{\displaystyle CH_3}{\underset{\displaystyle}{|}} \\
C_6H_5 - S \overset{*}{-} C - D \\
\overset{\displaystyle}{\underset{\displaystyle C_6H_5}{}}
\end{array}
$$

94

$$
\begin{array}{c}
\overset{\displaystyle CH_3}{\underset{\displaystyle}{|}} \\
KO_3S \overset{*}{-} C - D \\
\overset{\displaystyle}{\underset{\displaystyle C_6H_{13}\text{-}\underline{n}}{}}
\end{array}
$$

95

$$
\begin{array}{c}
\overset{\displaystyle O}{\underset{\displaystyle\|}{}} \quad \overset{\displaystyle CH_3}{\underset{\displaystyle}{|}} \\
C_6H_5 - P \overset{*}{-} C - H\,(D) \\
\overset{\displaystyle}{\underset{\displaystyle C_6H_5}{}} \quad \overset{\displaystyle}{\underset{\displaystyle C_6H_{13}\text{-}\underline{n}}{}}
\end{array}
$$

96

$$
\begin{array}{c}
\overset{\displaystyle O}{\underset{\displaystyle\|}{}} \quad \overset{\displaystyle CH_3}{\underset{\displaystyle}{|}} \\
EtO - P \overset{*}{-} C - H \\
\overset{\displaystyle}{\underset{\displaystyle EtO}{}} \quad \overset{\displaystyle}{\underset{\displaystyle C_6H_{13}\text{-}\underline{n}}{}}
\end{array}
$$

97

A number of ingenious systems were designed and studied by Corey et al.[91–93] and Bordwell et al.[94,95] to answer the question of the preferred geometry of the α-sulfonylcarbanion. Calculations of its structure were also performed by Wolfe et al.[96] The combined results pointed to a preferred pyramidal geometry for the ion existing in conformation 98. The rate of pyramidal inversion appeared to be fast relative to the rate of rotation around the C–S bond, at least partly because of the bulk of the substituents attached to both carbon and sulfur.[95] The same geometry probably applies to the class A carbanions just mentioned. Although we had the satisfaction of first reporting that the α-sulfonylcarbanion was asymmetric,[62] Corey and Bordwell were the "finalists" regarding its structure. Chemical research contests are different from tennis regarding first versus last shots.

98

Nothing definite can be said as to the geometries of the class B carbanions. The complete absence of k/k_α values less than unity for all eight d-orbital-containing systems studied suggests the carbanions to be all pyramidal but with different rates of pyramidal inversion and of conformational reorganizations, depending on substituents attached to sulfur or phosphorus.

Physical organic investigations of the type just described appeal to me partly because reaction mechanisms can never be fully settled. Their study engenders interesting, often pleasant, and always stimulating controversy. The aim of science is possibly the reduction of the mysterious to the rational. The purely rational, however, leaves me with a feeling resembling that of a good meal already eaten, rather than a feeling of eating a meal with several good courses yet to be served.

Dimethyl Sulfoxide Potentiation of Alkoxide Basicity (1959–1962)

One of the delights of organic chemical research is the serendipitous observations that turn out to have important general utility. In 1959, I suggested to J. L. Mateos that we examine the effect of dimethyl sulfoxide as solvent on the stereochemical course of the cleavage of **99** to **61**.[51] To our surprise, even the second and third times the experiments were performed, the reaction occurred readily at 25 °C in $(CH_3)_2SO$–t-C_4H_9OK, whereas in $O(CH_2CH_2)_2O$–t-C_4H_9OK, the reaction was slower even at 125 °C. In similar cleavages of **100** to **59** in $(CH_3)_2SO$ as solvent, comparable rates were observed with t-C_4H_9OK at 25 °C, t-C_4H_9ONa at 75 °C, and t-C_4H_9OLi at 170 °C. We concluded that the chemical reactivities of RO^-, $RO^- \cdots HOR$, RO^-K^+, RO^-Na^+, and RO^-Li^+ were all quite different. It appeared likely that ROK in $(CH_3)_2SO$ was a much more powerful base than had been previously thought possible for a metal alkoxide.

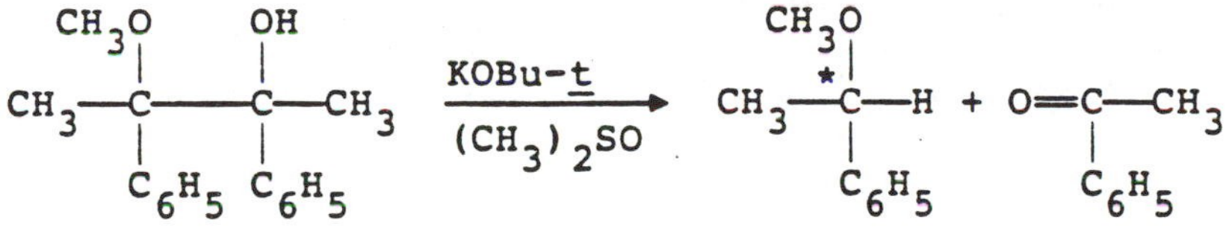

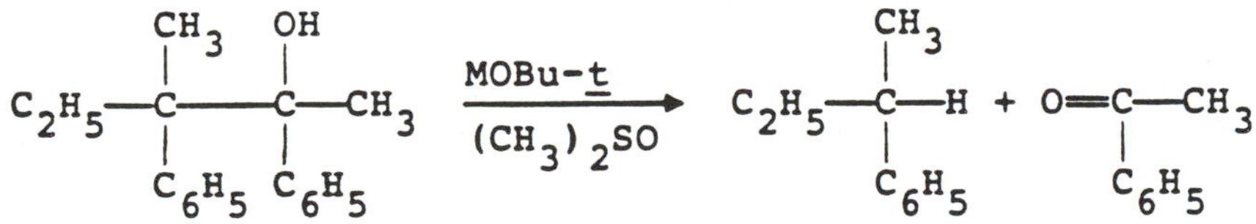

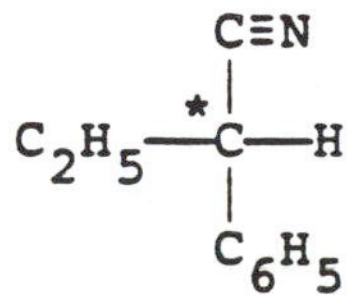

This expectation was confirmed in studies reported in 1960 by B. Rickborn, G. R. Knox, and me,[97] in which the rates of racemization of **67** catalyzed by $KOCH_3$ were measured in CH_3OH, $(CH_3)_2SO$, and mixtures of the two solvents. *The relative rates at 25 °C in pure $(CH_3)_2SO$ turned out to be $\sim10^9$ times faster than in pure CH_3OH.* Another dramatic example of this solvent effect, reported in the same paper, was the conversion at 25 °C of C_6H_5Br to $(CH_3)_3COC_6H_5$ in 86% yield when treated with t-C_4H_9OK in $(CH_3)_2SO$. The same reaction proceeded more slowly, in poorer yield at 175 °C, in $(CH_3)_3COH$. In 1962, it was demonstrated that the kinetic activity of CH_3OK ions in $(CH_3)_2SO$ had its thermodynamic counterpart.[98] A plot of $-\log k_2$ for racemization of **58** in various $(CH_3)_2SO\text{--}CH_3OH\text{--}CH_3OK$ mixtures against H^- (a measure of pK_a values of weak acids) was linear over a 10^6 change in k_2 and over a 10^7 change in dissociation constants for weak acids.[99] These early discoveries anticipated the widespread applications of $(CH_3)_2SO\text{--}ROK$ to syntheses, to mechanistic studies, and to the establishment of acidity scales of carbon acids.

Ever since the classic work of Conant and Wheland,[100] and McEwen,[101] the arrangement of carbon acids of all types on an acidity scale became a focal point for correlations between structure and reactivity. As a result of our use of stereochemical probes for carbanion structure, we needed pK_a estimates of a variety of carbon acids. In our 1965 book on carbanions,[47] we suggested a pK_a scale for carbon acids that was an amalgam of the work of many investigators, particularly that of McEwen, Streitwieser, Applequist, and Dessy. Our MSAD pK_a scale was named accordingly.

The scale ran from fluoradene ($pK_a \sim 11$) to cyclohexane ($pK_a \sim 45$). With this suggested scale came the qualifying statement, "Further work will undoubtedly lead to modifications of this scale, particularly at the upper end." The subsequent fundamental investigations of A. Streitwieser working with cyclohexylamine and F. G. Bordwell with $(CH_3)_2SO$ demonstrated this statement to be correct. The MSAD scale served two important purposes: (1) For a period of about 10 years, it was the most extended scale available and was much quoted. (2) In putting together the scale, we made outrageous assumptions that invited discreditation. We were not disappointed in this expectation.

Host—Guest Complexation (1970–1986)

Few scientists acquainted with the chemistry of biological systems at the molecular level can avoid being inspired. Three-and-a-half billion years of evolution have produced chemical compounds exquisitely organized to accomplish the most complicated and delicate of tasks. Many organic chemists viewing the cornucopia of structures of the enzyme systems, the nucleic acids, and the immune systems as they emerged over the years must have dreamed of designing and synthesizing compounds that would imitate some of their working features. I had this ambition as early as the late 1950s and recognized then that the study of highly structured complexes had to be the focus of research of this kind. Our investigations of the π complexes of the larger [$m.n$]paracyclophanes with $(NC)_2C=C(CN)_2$ led us to envision structures in which the π acid was sandwiched by the two benzene rings. However, no intercalated structures were observed.[35,36]

In 1967, the first papers of Pedersen[102,103] reported that crown ethers were capable of binding the alkali metal ions to generate highly structured complexes. I immediately recognized the crown ethers as an entrée into a general field of highly structured complexes. The papers in 1969 on the design, synthesis, and binding properties of the cryptands by J.-M. Lehn, J.-P. Sauvage, and B. Dietrich[104,105] further demonstrated the opportunities and attractions of this field. From 1968 on, I tried to interest graduate students in synthesizing *chiral* crown ethers, but I was unsuccessful. In 1970, I insisted that several of my postdoctoral co-workers enter the field, and during 1973 we published five Communications on the subject.[106–110] Graduate student resistance to accepting such research problems "melted." In 1974, I published, with Jane M. Cram, a general article entitled "Host–Guest Chemistry", in *Science*,[111] that defined our general approach to this research.

From the beginning, I visualized a field in which structurally organized host compounds containing cooperating functional groups would form highly structured complexes with guests. Guests might be other organic compounds or ions, inorganic ions, or salts of organic compounds. Many developments in the 1960s made this research timely and attractive:

Big Brother Cram and his good friend, Teruaki Mukaiyama in Cram's office, 1979.

1. Crystal structure determinations of compounds of molecular weights in the high hundreds were becoming possible.

2. Nuclear magnetic resonance techniques provided windows for gathering structural information on solution structures of organic compounds.

3. Corey–Pauling–Koltun molecular models based on crystal structures of peptides and proteins became available.

4. Gel permeation chromatography was coming into use and allowed compounds to be easily separated according to their sizes. Because many of the hosts were oligomeric and had molecular weights of 1000 ± 500, this technique was of great importance in isolating and recognizing desired products.

5. Many new organic reactions, reagents, and protective groups were being developed.

6. Enzyme structures and their modes of action were being elucidated at a rate that suggested an analogous synthetic organic counterpart should be invented to parallel the discoveries of the biological chemistry as they emerged.

7. Research of the field of host–guest complexation provided a fine vehicle for teaching graduate students and postdoctoral scholars how designs and techniques merge in modern organic chemical research.

By the time I was 50 years old, my co-workers' results and their interpretations had provided me with an education appropriate to this field. I personally needed challenges of a new nature to freshen my spirit of inquiry. Compounds had to be designed for interesting purposes, then their synthesis and test systems had to be designed, and the concepts had to be reduced to practice.

My third graduate student (of 1948 vintage) was Frederick M. Hawthorne, who is now one of my prized colleagues. While having lunch with him in about 1969, just before we changed our research field, he made the following offhand remark: "Don, I was just looking at the last issue of the *Journal of the American Chemical Society* and saw four papers there of yours. I looked at them, but not too carefully, since it seemed to me I had seen them already." This statement was made quizzically and with apparent innocence. When a former co-worker who is a fine scientist and a good friend makes a remark like this, it is high time you change your research field.

A new field requires new terms, which, if properly defined, greatly facilitate the reasoning by analogy on which research thrives. In our first full-length paper in the series, *Host–Guest Complexation*,[112] we defined the following terms and identified the central concepts as follows:

> Complexes are composed of two or more molecules or ions held together in unique structural relationships by electrostatic forces other than those of full covalent bonds . . . . Molecular complexes are usually held together by hydrogen bonding, by ion pairing, by π acid to π base interactions, by metal to ligand binding, by van der Waals attractive forces, by solvent reorganizing, and by partially made and broken covalent bonds (transition states) . . . . High structural organization is usually produced only through multiple binding sites . . . . A highly structured molecular complex is composed of at least one host and one guest component . . . . A host–guest relationship involves a complementary stereoelectronic arrangement of binding sites in host and guest . . . . The host component is defined as an organic molecule or ion whose *binding sites* converge in the complex . . . . The guest component is defined as any molecule or ion whose *binding sites* diverge in the complex.[112]

Thus *hosts* are synthetic counterparts of the receptor sites of biological chemistry, and *guests*, the counterparts of substrates, inhibitors, or cofac-

Donald J. Cram with R. B. Woodward holding the microphone, circa 1968, in Belgium at an IUPAC meeting.

tors of biological chemistry. In the period from 1973 to 1986, these terms and concepts gained broad international acceptance.

We introduced these terms because they were much needed, but only after they had passed "Cram's test." Each of my co-workers was asked to recite the proposed new terms at those times of day when they were most depressed. At the end of the week, if they could still hear the words without total revulsion, the terms passed the test. The consensus of my research group was that the terms could be tolerated, particularly because no better ones had occurred to them.

The Molecular Model Design of Complexes: Crystal Structure Connections

Corey–Pauling–Koltun (CPK) molecular models were developed under the auspices of the National Institutes of Health, the National Science Foundation, and the American Society of Biological Chemists "for constructing macromolecules of biological interest." The models are based on crystal structures of biologically important compounds.[113] I envisioned these models as connecting the biotic world of evolutionary

Donald J. Cram singing and playing the guitar in Kyoto, Japan, before several hundred guests during a chemical meeting in 1975, with Yasuhida Yukawa holding the microphone.

chemistry and the abiotic—conceptual world of the organic chemist, as indicated in Chart I. From the beginning, these models served as a compass in an uncharted sea of potential target complexes. Accordingly, I started this research by spending hundreds of hours building CPK models of potential complexes and grading them as potential research targets in terms of probable scientific yield (value of results versus effort expended). Complexes were then prepared, and their crystal structures determined to assess the correspondence between what was anticipated by model examination and what was experimentally observed. This comparison would not have been possible without collaboration between my research group and crystallographers. By the end of 1986, K. N. Trueblood, C. B. Knobler, E. F. Maverick, and I. Goldberg,[114] working at UCLA, had determined the crystal structures of over 50 complexes, thus making an invaluable contribution to the progress of host—guest complexation chemistry.

We started this research with faith in our predictive powers based on CPK molecular modeling. The crystal structures turned our faith into confidence. In all but 2 out of over 50 complexes prepared at UCLA by 1986, predictions of their structures based on molecular models were borne out by crystal structures. Chart II compares predicted with observed structures for complexes designed and prepared in the early 1970s. In some cases, publications of the crystal structures occurred many years after the complexes were prepared, or after they had been determined.

In the early 1970s, after designing but before preparing many of the complexes of Chart II, I gave many seminars and had many discussions composed largely of predictions of structures and expected properties of such complexes. These talks were rich in speculation and poor in results. They featured heavy, space-occupying CPK molecular models of numerous types of complexes. These were my traveling companions. They elicited much curiosity on the part of customs inspectors. My talks were dignified by the "grand biotic—abiotic analogy." About 1972, my long-time friend, John D. Roberts, dryly pointed to my combined bravery and optimism in giving many seminars on "work not yet done!" In 1985, A. Eschenmoser reminded me of a chat we had in 1972 in Switzerland, where, models in hand, I had tried to substitute excitement and enthusiasm for experimental results in convincing him of the virtues of my abiotic models. He recalled that, after that 1972 conversation, he had thought (but had been kind enough not to say) that I should have my head examined. He then added, "You must take great satisfaction in seeing these dreams become reality." Remarks such as these, made by people we admire, are one of the important kinds of currencies in which we investigators are paid. As our research matured, my seminars fastened on results in-hand.

Chart I. Molecular design, molecular model, and crystal structure connections

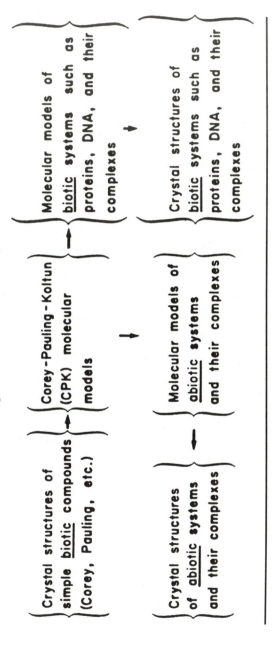

Complexes **101–103** are intramolecular, demonstrating the predictability of attractive hydrogen-bonding interactions between $CO_2H \cdot O(CH_2)_2$, $CO_2H \cdot NC_5H_5$, and $CO_2H:HO_2C$, respectively. All three complexes also showed attractions of the $HO_2C \cdots O(CH_2)_2$ type, whose covalent counterparts are the *ortho* intermediates of transacylation reactions.[114–116] Complex **104** demonstrates *tripod binding* of a primary amine salt to a crown ether containing a lipophilizing naphtho group.[117] Complex **105** shows that a pyridine could be substituted for a CH_2OCH_2 group in a *corand* (general term for crown-like hosts) without disturbing the tripod structure of the *coraplex* (general term for complexes of corands).[118] The $N^+H \cdot NC_5H_5$ interaction in **105** is preferred over an alternate structure containing only $N^+H \cdot O(CH_2)_2$ binding interactions. Complex **106** contains a chiral host. One methyl of the host is inserted between the two methyls of the guest.[119] Unlike the other coraplexes, **107** contains a $CO_2^- \cdot HN^+$ binding site.[115] The host of **108** is a member of a family of ligating systems, which we named *hemispherands*, whose three anisyl groups preorganize themselves for binding during their syntheses. Hemispheraplex **108** contains a bifurcated hydrogen bond as a binding site $N^+H:(OCH_3)_2$.[114,120] Complexes **104–108** all contain the $(CH_3)_3CNH_3^+$ ion as guest whose C–N bond is essentially normal to the plane of the three or four heteroatoms involved in hydrogen bonding in each of the complexes, as predicted by model examination.[114]

Complex **109** provided the first example of two sets of binding sites in both host and guest that were positioned to be complementary to one another.[121] Coraplex **110** illustrates how two sets of hinged ligating sites can close like jaws on a single metal ion, guest K^+.[121] In **111**, two interlocking corand strands cooperatively bind K^+ as guest.[121] Complex **112** combines a corand binding site with two complementarily placed side arms terminated with $P \rightarrow O$ binding sites. All eight oxygens of the host contact the K^+ guest in a roughly hexagonal bipyramidal arrangement.[121,122]

Several distinguishing features are found in **113**.[114,121] The guest is a "neutral" water molecule donating two hydrogen bonds to two ether oxygens while accepting a hydrogen bond from a phenolic hydroxyl of the host. Coraplex **114** provided the first demonstration that corands could complex aryldiazonium salts and stabilize them,[112,123,124] an observation seminal to more extensive investigations by G. W. Gokel and R. A. Bartsch. We provided chemical evidence[112] for the "plugged in" structure of **114** long before B. Haymore[124] determined its crystal structure.

Our coraplex, **115**, in which a guanidinium ion guest donates six hydrogen bonds to six strategically placed oxygens, was reported in 1977.[112] Its crystal structure, reported in 1983 by S. Harkema et al.[125]

Chart II. Molecular model structures

Molecular Model Structures   *Crystal Structures*

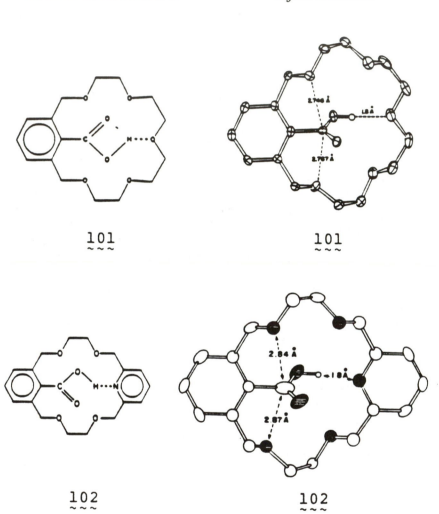

101   101

102   102

compared with crystal structures

Molecular Model Structures   Crystal Structures

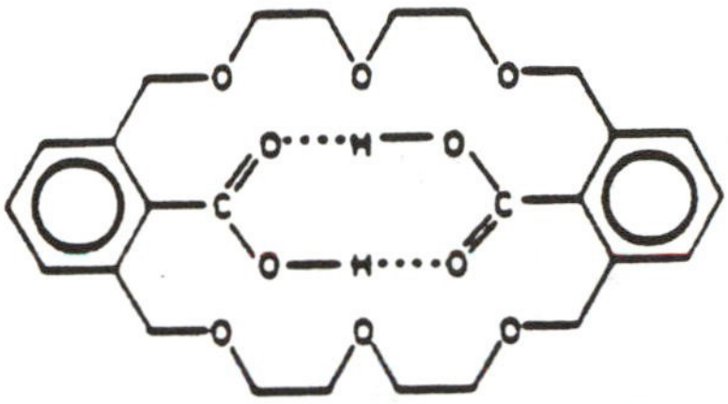

103

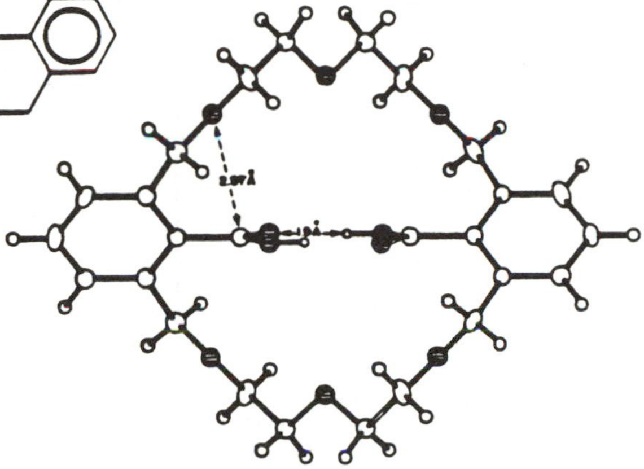

103

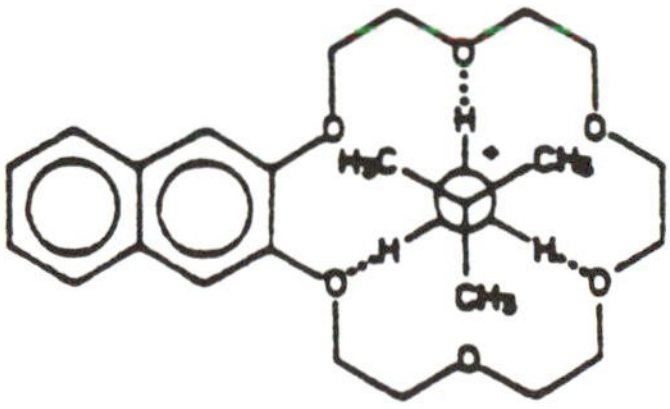

104

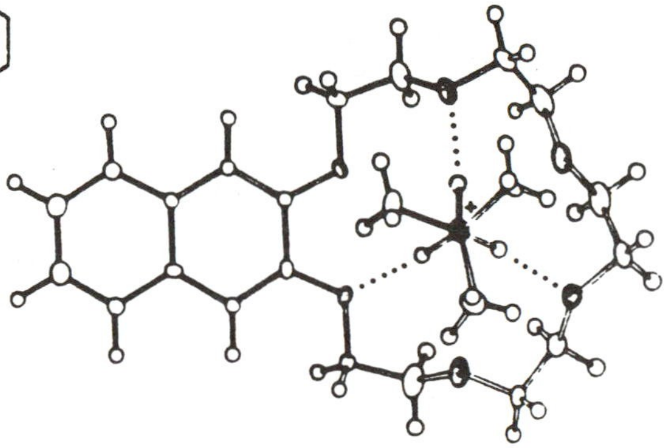

104

Continued on next page

Chart II. Molecular model structures

Molecular Model Structures   *Crystal Structures*

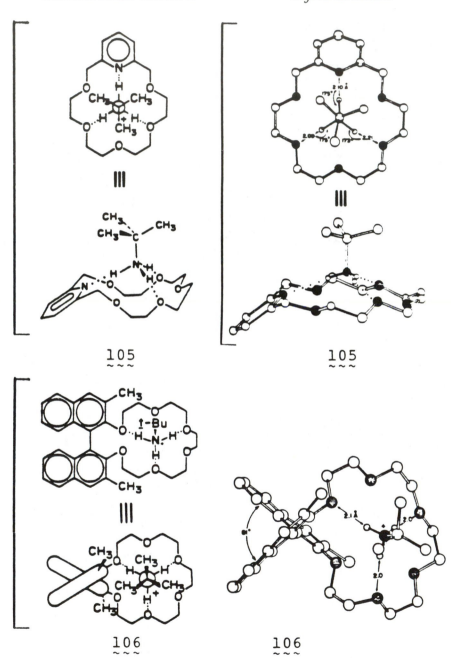

III   III

105   105

III

106   106

compared with crystal structures—Continued

Molecular Model Structures   *Crystal Structures*

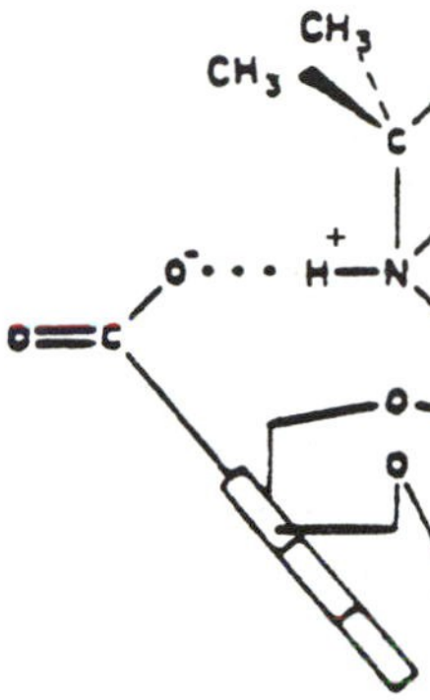

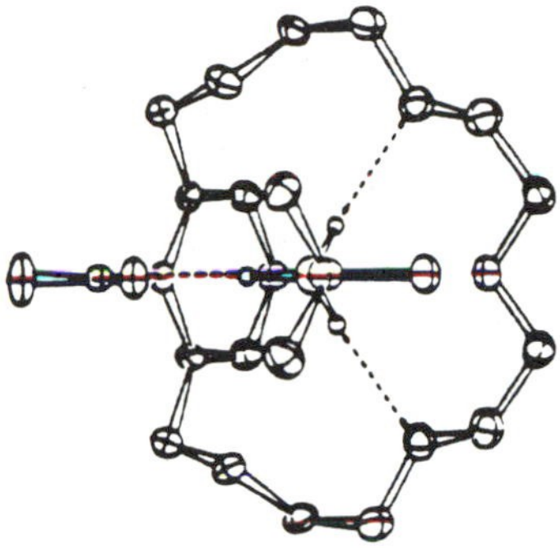

107

107

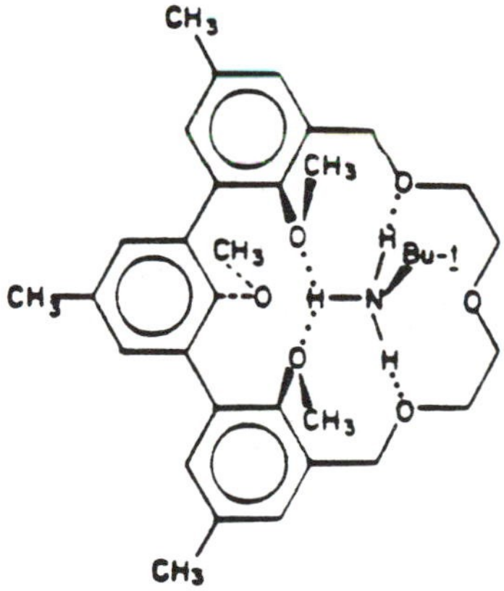

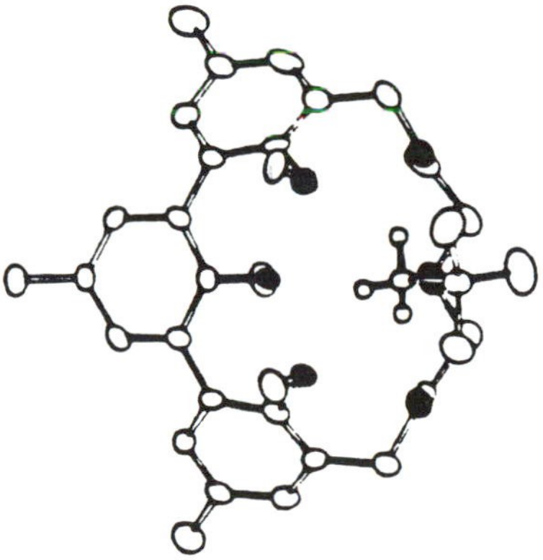

108

108

Continued on next page

Chart II. Molecular model structures

Molecular Model Structures   *Crystal Structures*

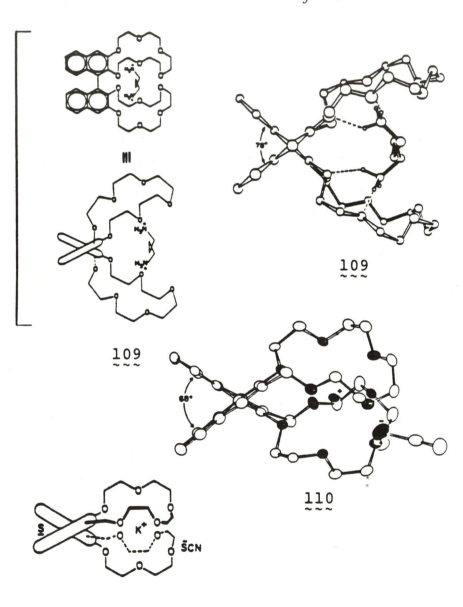

111

109

109

110

110

compared with crystal structures—Continued

Molecular Model Structures   *Crystal Structures*

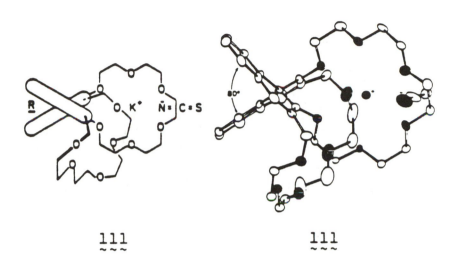

$\underset{\sim\sim\sim}{111}$ $\underset{\sim\sim\sim}{111}$

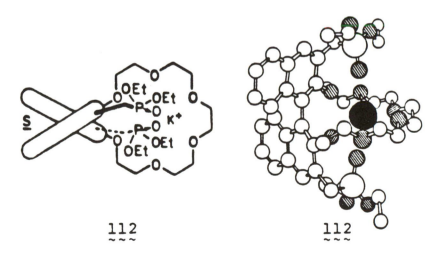

$\underset{\sim\sim\sim}{112}$ $\underset{\sim\sim\sim}{112}$

Continued on next page

Chart II. Molecular model structures

Molecular Model Structures   *Crystal Structures*

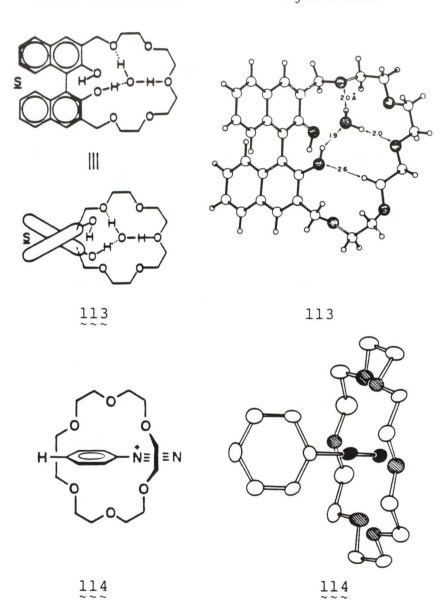

113   113

114   114

compared with crystal structures—Continued

Molecular Model Structures   Crystal Structures

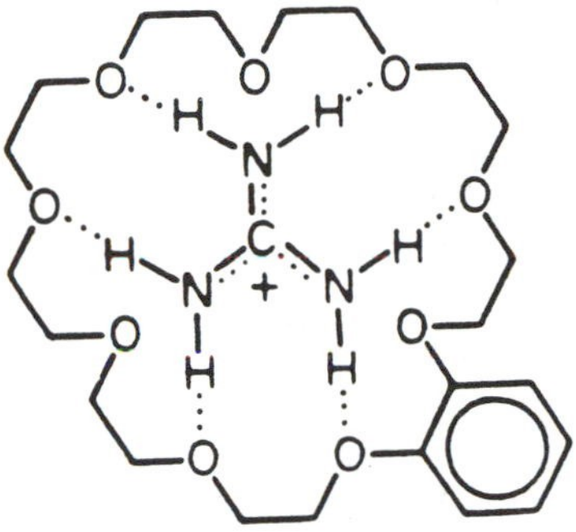

115 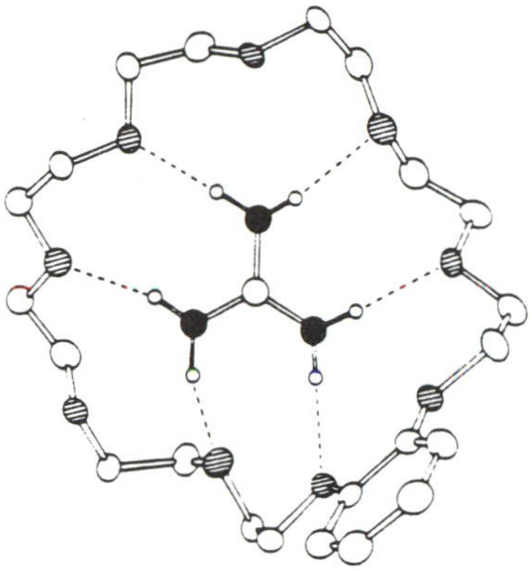   115

confirmed the anticipated wreathlike structure. These complexes use a wide variety of functional groups that act as binding sites and hydrocarbon groups that provide scaffolds, spacers, and steric barriers to the complexing partners. I designed each complex by manipulating molecular models. The complexes were prepared by my co-workers, and the crystal structures were determined by my colleagues. This body of results provided the first demonstration that a wide variety of complexes could be designed by rational means through the use of CPK molecular models and a few physical organic concepts. These results provided a foundation for design of the more sophisticated systems that followed.

Inevitably, the complexes acquired self-explanatory class names based on their general shapes. Complexes 104–108 are referred to as *perching*; complexes 109, 111, and 113–115 as *nesting*; and complexes 110 and 112 as *capsular*. Many co-workers from the early 1970 period entered academic work, such as E. P. Kyba, K. Koga, R. C. Helgeson, G. W. Gokel, L. R. Sousa, M. Newcomb, G. R. Weisman, T. Kaneda, and D. M. Walba.

Chiral Recognition in Complexation

Structural recognition in complexation is one of the most important means by which the receptor sites of genes, enzymes, and immune systems regulate chemical traffic in the evolutionary, biotic world. In our first ventures into designed host–guest complexation chemistry in 1973,[106,107] we incorporated 1,1-binaphthyl units into corands to study their potential chiral recognition in complexation of amino acid and ester salts. We were careful to design systems that contained at least one C_2 axis of symmetry, a tactic that made the hosts nonsided with respect to perching guests. Structural refinements of the early systems finally led to host **116** as exhibiting the highest chiral recognition. A solution of (*R,R*)-**116** in CDCl₃ at 0 °C was used to extract D₂O solutions of racemic amino acid or ester salts. As predicted in advance by CPK

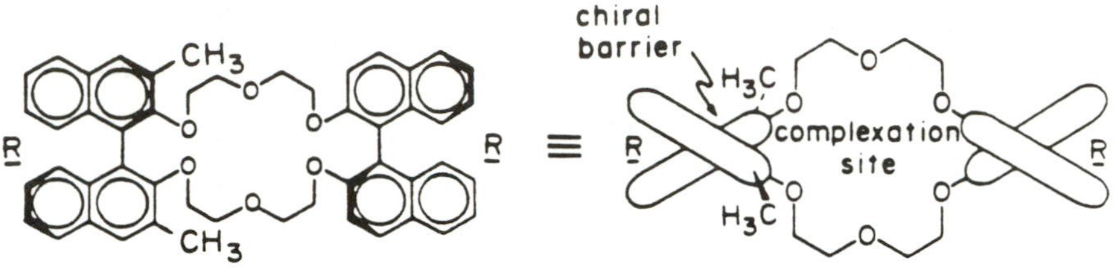

$$\underset{\sim\sim\sim}{116}$$

molecular models, the (D)-enantiomers were extracted preferentially into the organic layer. Chiral recognition factors ranged from a high of 31 for C₆H₅CH(CO₂CH₃)NH₃PF₆ to a low of 2.3 with CH₃CH(CO₂H)NH₃ClO₄. These factors represent free-energy differences between diastereomeric complexes of 1.9 and 0.42 kcal mol⁻¹, respectively, with other amino acids and ester salt guests ranging between these values. We interpreted these results in terms of the complementarity between host and guest of the (*R,R*)-(D) configurations as visualized in coraplex **117**, and the lack of complementarity in those of the (*R,R*)-(L) configurations, which were designed not to form.[126,127]

In work with M. Newcomb, J. L. Toner, and R. C. Helgeson,[128] an amino acid and ester resolving machine was designed, built, and tested;

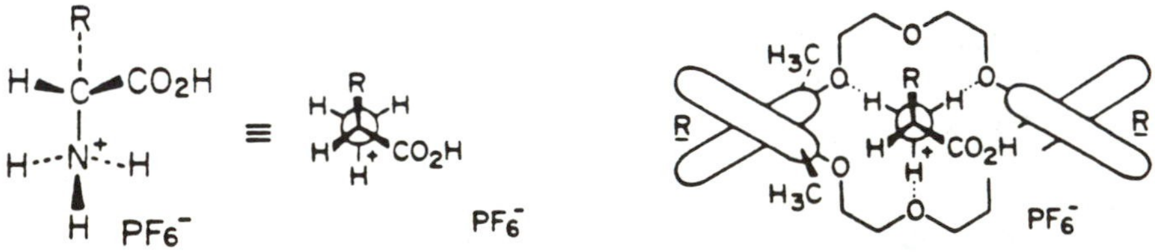

$$\underset{\sim\sim\sim}{117}$$

STABLER COMPLEX

it is pictured in Figure 1. The machine used chiral recognition in transport of amino acid or ester salts through lipophilic liquid membranes. From the central reservoir of the W-tube containing an aqueous solution of racemic salt, the (L)-enantiomer was picked up by (S,S)-**116** in the left chloroform reservoir and delivered to the left aqueous layer, while the (D)-enantiomer was transported by (R,R)-**116** in the right chloroform reservoir and delivered to the right aqueous layer. The thermodynamic driving force for the machine's operation involved exchange of an energy-lowering entropy of dilution of each enantiomer for an energy-lowering entropy of mixing. To maintain the concentration gradients down which the enantiomers traveled in each arm of the W-tube, fresh racemic guest was continuously added to the central reservoir, and (L)- and (D)-$C_6H_5CH(CO_2CH_3)NH_3PF_6$ of 86–90% ee (enantiomeric excess) were continuously removed from the left and right aqueous reservoirs, respectively.[128]

In another experiment, G. D. Y. Sogah[129] covalently attached the working part of the (R,R)-**116** at a remote position to a macroreticular resin (polystyrene–divinylbenzene) to give immobilized host of ~18,000 mass units per average active site. This material was used to give complete enantiomeric resolution of several amino acid salts. The behavior in the chromatographic resolution paralleled that observed in the extraction and transport experiments and was useful both analytically and

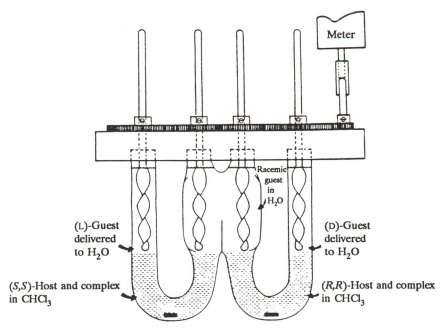

Figure 1. Resolving machine.

Donald J. Cram, Yasuhida Yukawa, and Shigeru Oae in Kyoto in 1975.

preparatively. Separation factors ranged from 26 to 1.4, the complexes of the (R,R)-(D)- or (S,S)-(L)-configurations always being the more stable. The structure envisioned for the more stable complex is formulated in **118**.

In work with D. S. Lingenfelter in 1981,[130] the chiral barriers of the 1,1-binaphthyl module were extended by attachment of phenyl groups in the 2,2-positions, as in **119**. This host is perhaps more useful for amino acid and ester resolution because it contains only one chiral element. It provided chiral recognition factors that ranged from as high as 22 to as low as 4 toward 12 different amino acids and esters. In all cases, the (R)-(D)- or enantiomeric (S)-(L)-complexes were more stable

$$\underset{\sim\sim\sim}{\underline{118}}$$

than the diastereomeric (R)-(L)- or (S)-(D)-complexes. The same generalization applied to complexes whose hosts contained nine different A groups of general structures **120** and **121**, and two different R' groups.

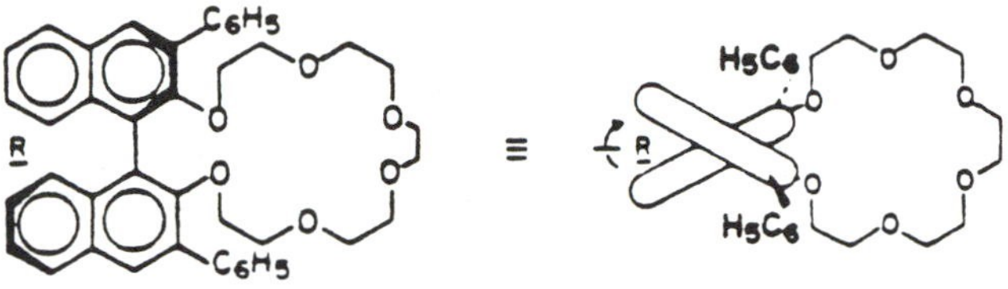

119

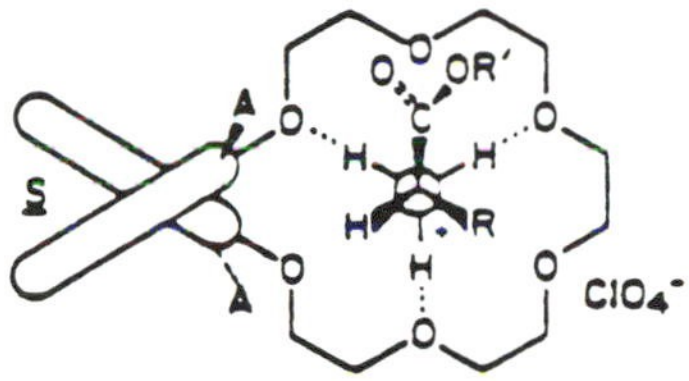

120

(S)(L)–more stable

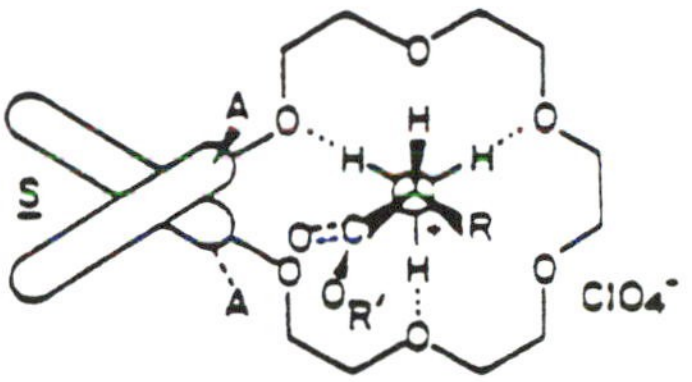

121

(S)(D)–less stable

Catalysis by Chiral Hosts

Alkaloids had been shown to be turnover catalysts for the Michael addition reactions by research carried out by G. Bergson and B. Langstrom,[131] and later by H. Wynberg and K. Hermann.[132] Modest asymmetric induction by these chiral amine catalysts had been observed. Molecular model examination of contact potassium carbanionide ion pairs complexed to host (R,R)-**116** appeared to provide a highly asymmetric environment for the planar prochiral carbanion **122**. In the envisioned complex **124**, the carbanion becomes "sided", and electrophiles, such as 3-butenone, approach the carbanion only from the unshielded side to produce **123**. We predicted, in advance of experiment, the dominant configuration of the product to be (S)-**123**. In practice, G. D. Y. Sogah and I observed[133] (1981) high catalytic turnover, very high asymmetric induction (ee, ~100%), and the anticipated configuration (S) for **123**. In the absence of host, no reaction occurs because of the insolubility of **122** in the medium (toluene).

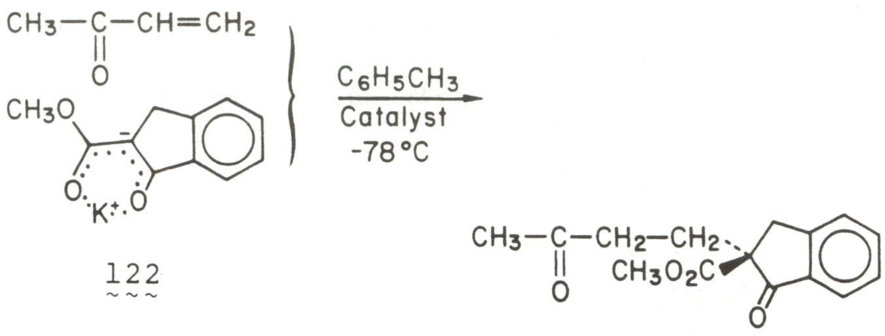

$CH_3-C(-O)-CH=CH_2$

CH_3O

122

$\xrightarrow[{-78\,°C}]{\begin{array}{c} C_6H_5CH_3 \\ \hline Catalyst \end{array}}$

$CH_3-C(=O)-CH_2-CH_2$
CH_3O_2C

(**S**)-**123**

$CH_2=CH-C(=O)-CH_3$

H_3C R K^+ R H_3C

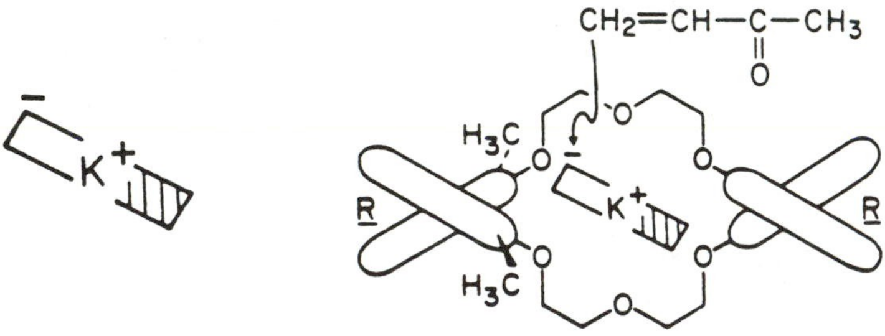

122   **124**

    In a study with J.-P. Mazaleyrat[134] (1981), we investigated the possibility that **125** might act as a stoichiometric catalyst for asymmetric induction in the addition of alkyl- and aryllithium reagents to aldehydes. We envisioned that the direction of the chiral bias in the product would be controlled by a transition state suggested by formula **126**. The effective D_2 symmetry of **125** allowed **126** to be composed in a variety of ways that avoided competing diastereomeric transition states. In practice, **125** provided asymmetric induction for the reactions of five different sets of reacting partners to give optical yields that ranged from 36 to 94% ee. In all cases, **126** predicted the observed configurations of the products. Molecular models of **126** suggested that the optical yields should all have been close to 100% ee. Control experiments showed

$$R'-\overset{\overset{\displaystyle H}{|}}{C}=O \quad + \quad R-Li \cdot Host \quad \xrightarrow[\substack{-120°C \\ 2)\ H^+}]{1)\ Et_2O} \quad R'-\overset{\overset{\displaystyle H}{|}}{\underset{\underset{\displaystyle OH}{|}}{C}}-R \quad + \quad Host$$

125   126

that noncatalyzed reactions competed with the catalyzed reaction to lower the optical yields of products. Both D. Seebach et al.[135] and T. Mukaiyama et al.[136] had previously used asymmetric catalysts based on natural products to get optical yields of 15 to 92% in similar reactions.

Because anionic polymerization of methacrylate esters is a special example of the Michael addition reaction, we anticipated that hosts 116 and 127 complexed with metal ion bases might act as chiral catalysts to

Teruaki Mukaiyama and Donald J. Cram in Cram's office, examining models, 1982. See also pp 51 and 73.

"stamp out" asymmetric units to give optically active isotactic polymers. The carbanion ion-paired to the host-bound cation is present at the growing end of the chain; thus, any configurational "mistake" would tend to be corrected rather than be perpetuated in the successive generation of asymmetric centers. While this work with G. D. Y. Sogah was in progress, Yuki et al.[137] reported that (−)-sparteine–C_4H_9Li-initiated polymerization of trityl methacrylate gave insoluble, optically stable isotactic polymer whose high optical rotation was attributed to helicity.[137] Accordingly, we included **125** in our study. The only other known synthetic polymers that are helically chiral are the polyalkyl isocyanides, which show configurational stability (W. Drenth and R. J. M. Nolte).[138]

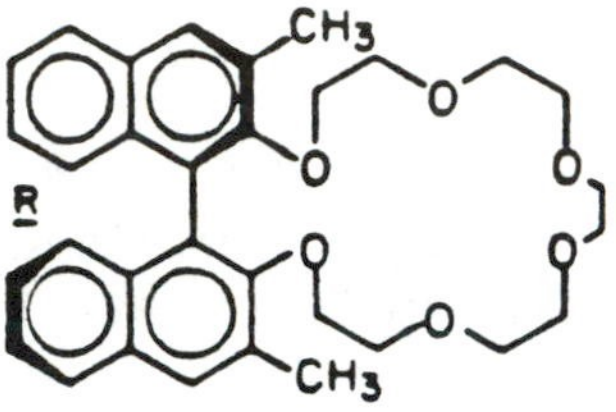

127

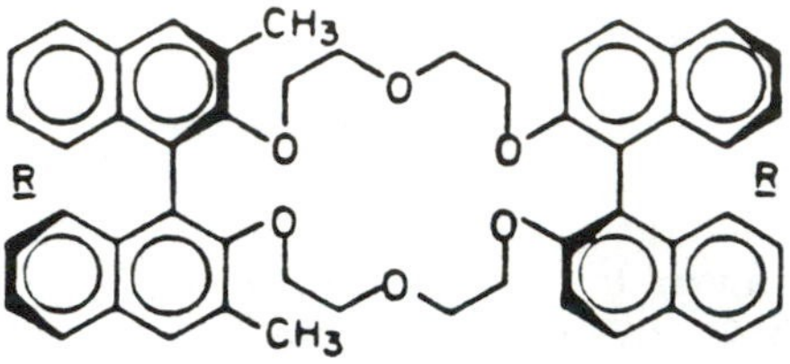

116

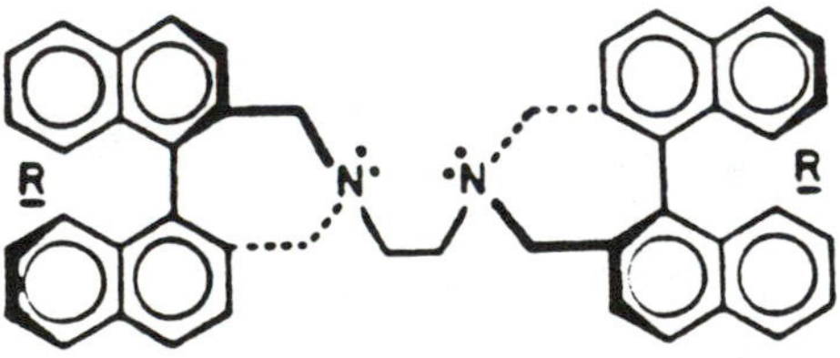

125

We used optically pure **116**·t-C_4H_9OK, **127**·t-C_4H_9OK, and **125**·t-C_4H_9OLi as initiators of polymerization of methyl, t-butyl, and benzyl methacrylate at −78 °C in 95% toluene–tetrahydrofuran to give polymers of 80–90% isotacticity. The polymers were purified rapidly by gel permeation chromatography to give $\overline{M}_n$ values of about 1100–2100 mass units and, initially, high optical rotations. In solution, the poly(methyl methacrylate) and poly(t-butyl methacrylate) mutarotated at ambient temperature over periods of many hours to give material of optical rotations close to zero. We concluded that the high rotations were due to helicity of the polymers induced by the chiral cavities of

Teruaki Mukaiyama, Donald J. Cram, and Jane Cram in Donald's office, 1984. Notice that the plant in the background and the breadth of the smiles have grown from 1979 to 1984. See also pp 51 and 71.

the catalyst in which each unit was added to the growing chain. The helices appeared to be thermodynamically unstable kinetic products of polymerization.

The patterns of relationships between the configurations of the catalysts employed and the signs of rotations of the helical polymers produced were coherent with the patterns observed for the simple Michael additions with **116** and **127** as catalysts. On the basis of such analogies, we suggested that **128**, **129**, and **130** (in which P is polymer) as representing models for the dominant transition states for the carbon–carbon bond-forming steps of the propagation reactions.

128

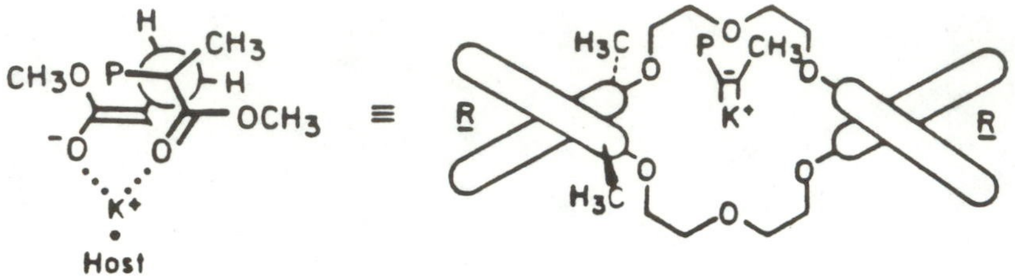

129

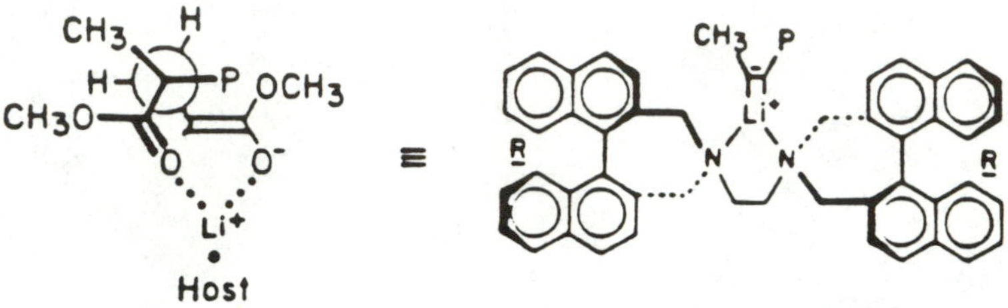

130

Formulas **131** and **132** represent nonhelical and helical poly(methyl methacrylate), respectively, whose formation was initiated with complexed t-C_4H_9OK. Models **128** and **130** predict the configurations of the asymmetric centers of the polymer and the counterclockwise helical configuration shown in **132**, whereas **129** predicts product of the opposite configuration about the asymmetric centers and the clockwise helical configuration.[139]

Spherands and the Principle of Preorganization

Crystal structures of corand **133**[140] and cryptand **135**[141,142] show that, in their uncomplexed state, they contain neither cavities nor convergently arranged binding sites. Guest ions, such as K^+, conformationally reorganize and desolvate corands and cryptands upon complexation, as shown by comparisons of the crystal structures of **133** with **134** and of **135** with **136**. In 1976, with the help of CPK models, I designed ligand system **137**, whose oxygens in CPK models appeared octahedrally arranged around an enforced spherical cavity complementary to Li^+ or Na^+ ions. We have given the family name, *spherand*, to completely preorganized ligand systems. Shortly thereafter, Takahiro Kaneda et al. started work on the synthesis of **137**, which was completed and

Colleagues at UCLA, about 1977, left to right, Fred Hawthorne, Tom Jacobs, Orville Chapman, Mike Jung, Donald J. Cram, John Gladysz, Frank Anet, and Larry Scott.

reported in 1979.[143] The crystal structures of spherand **137** and its corresponding spheraplexes **138** and **139** were reported as **140, 141,** and **142,** respectively in 1981.[144] As expected, the structure of **140** contains a hole lined with 24 electrons that are shielded from solvation by six aryl and six methyl groups. The snowflakelike structures of **140** and the ligand systems of spheraplexes **141** and **142** are almost identical. Thus, **137** represented the first ligand system that was designed and synthesized to be completely organized for complexation during its synthesis, rather than during its act of complexation.

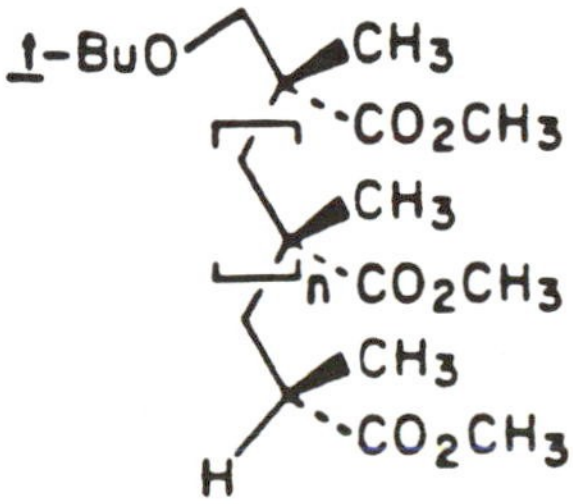

<u>t</u>-BuO CH_3
CO_2CH_3
CH_3
n CO_2CH_3
CH_3
CO_2CH_3
H

$\underset{\sim\,\sim}{131}$

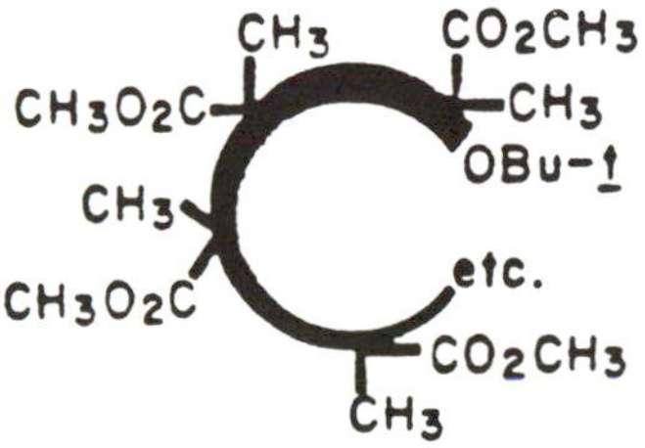

CH_3 CO_2CH_3
CH_3O_2C CH_3
OBu-$\underline{t}$
CH_3
CH_3O_2C etc.
CO_2CH_3
CH_3

$\underset{\sim\,\sim}{132}$

Donald J. Cram and Kurt Mislow in penguin suits on the occasion of receiving honorary degrees from the University of Uppsala, 1977.

The synthesis was much delayed because we had to invent a new ring-closing reaction to obtain the compound. When the synthesis was completed, Kaneda asked me how often I had considered abandoning the research problem. I replied, "The idea has never occurred to me." I then asked him how often he had thought of it. He replied, "at least once a week for 2 years!" I pointed to the great faith I had had in the viability of the problem and in his research ability. He, in turn, pointed to his faith in my faith. Endeavor has ever been thus!

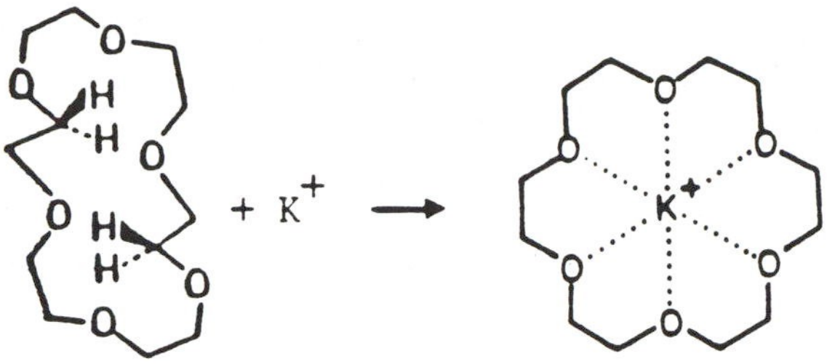

133

134

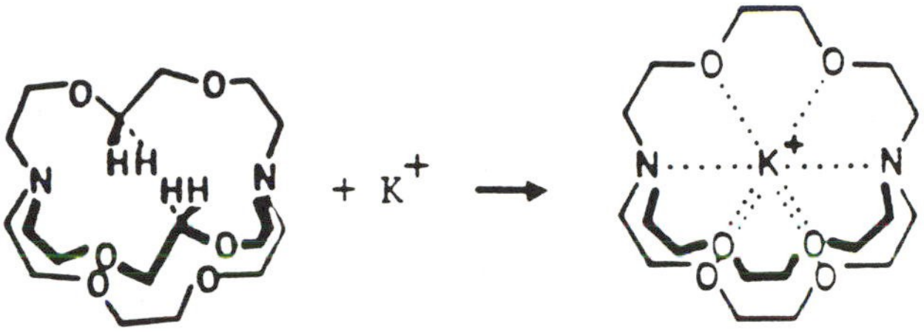

135

136

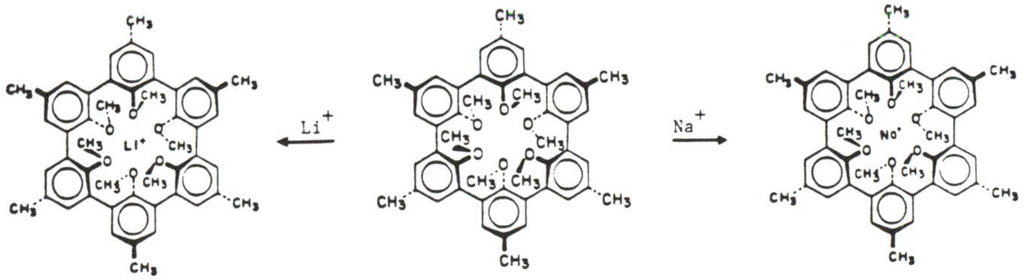

138   137   139

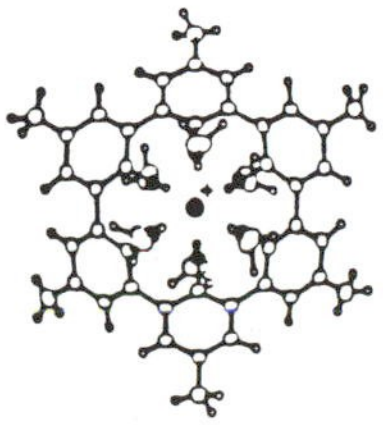

141 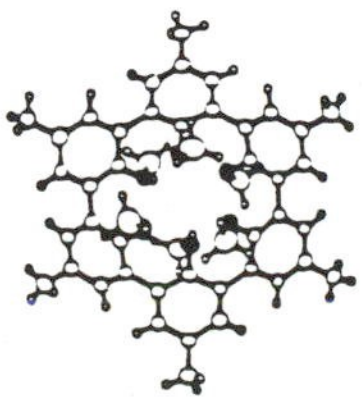   140 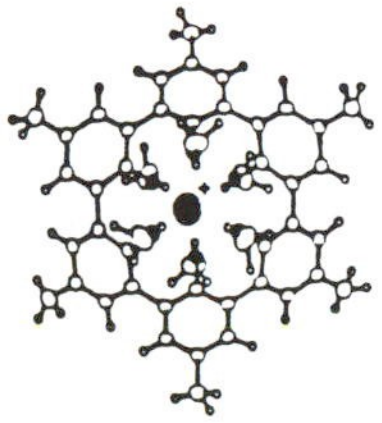   142

We developed[145] a means of determining the binding free energies of lipophilic hosts toward guest picrate (Pic) salts of Li^+, Na^+, K^+, Rb^+, Cs^+, $NH_4{}^+$, $CH_3NH_3{}^+$, and t-$C_4H_9NH_3{}^+$. The guest salts were distributed between $CDCl_3$ and D_2O at 25 °C in the presence and absence of host. From the results, K_a (mol^{-1}) and $-\Delta G°$ values (kcal mol^{-1}) were calculated (equation 10).

$$H + GPic \underset{k_{-1}}{\overset{k_1}{\rightleftharpoons}} H \cdot G \cdot Pic \tag{10a}$$

$$K_a = \frac{k_1}{k_{-1}} \tag{10b}$$

$$-\Delta G° = RT \ln K_a \tag{10c}$$

This method was rapid and convenient for obtaining $-\Delta G°$ values at 25 °C that ranged from about 6 to 16 kcal mol^{-1} in $CDCl_3$ saturated with D_2O.[145] Higher values (up to 22 kcal mol^{-1}) were obtained by equilibration experiments between complexes of known and unknown $-\Delta G°$ values.[146,147] Other values were determined from measured k_{-1} and k_1 values, all in the same medium at 25 °C.[146,147] Spherand 137 binds LiPic with $-\Delta G°$ values >23 kcal mol^{-1}, NaPic with 19.3 kcal mol^{-1}, and totally rejects the other standard ions, as well as a wide variety of other di- and trivalent ions.[147] The open-chain counterpart of spherand 137, podand 143, binds LiPic and NaPic with $-\Delta G° < 6$ kcal mol^{-1}.[148]

Podand 143 differs constitutionally from spherand 137 only in the sense that 143 contains two hydrogen atoms in place of one Ar—Ar bond in 137. The two hosts differ radically in their conformational structures and states of solvation. The spherand possesses a single conformation ideally arranged for binding Li^+ and Na^+. Its oxygens are deeply buried within a hydrocarbon shell. The orbitals of their unshared electron pairs are in a microenvironment whose dielectric properties are between those of a vacuum and a hydrocarbon. No solvent can approach these six oxygens that remain unsolvated. The free-energy cost of organizing the spherand into a single conformation and of desolvating its six oxygens was paid during its synthesis. Thus, spherand 137 is preorganized for binding.[149]

The podand, in principle, can exist in over 1000 conformations, only two of which can bind metal ions octahedrally. The free energy for organizing the podand into a binding conformation and desolvating its six oxygens must come out of its complexation free energy. Thus,

$$\underset{\sim\sim\sim}{143}$$

the podand is not preorganized for binding but is randomized to maximize the entropy of mixing of its conformers and to maximize the
attractions between the solvent and its molecular parts.

The difference in $-\Delta G^\circ$ values for spherand 137 and podand 143
for binding Li^+ is >17 kcal mol^{-1}, corresponding to a difference in K_a

*Celebration after the 29th IUPAC Congress in Cologne, Germany,
1983. From left to right, Heinz Staab, Jane Cram, Donald J. Cram,
Mrs. Vogel, and Emmanuel Vogel.*

of a factor of $>10^{12}$. The difference in $-\Delta G^\circ$ values for **137** and **143** for binding Na^+ is >13 kcal mol^{-1}, corresponding to a difference in K_a of a factor of $>10^{10}$. These differences are dramatically larger than any we have encountered that are associated with other effects on binding power toward alkali metal ion guests. We concluded[149] that *preorganization is a central determinant of binding power*. We formalized this conclusion in terms of what we call the *principle of preorganization*, which states that "the more highly hosts and guests are organized for binding and low solvation prior to their complexation, the more stable will be their complexes." Both enthalpic and entropic components are involved in preorganization, because solvation contains both components.[147] Furthermore, binding conformations are sometimes enthalpically rich. For example, the benzene rings in spherand **137** and spheraplexes **138** and **139** are somewhat folded from their normal planar structures to accommodate the spatial requirements of the six methoxyl groups.

As measured by its ability to hydrogen bond, the anisyl group is an intrinsically poor ligand.[150,151] We were not surprised that podand **143** is such a weak binder. That spherand **137** is such a strong binder provides an extreme example of the power of preorganization.

Families of hosts generally fall into the order of their listing in Chart III when arranged according to the $-\Delta G^\circ$ values with which they bind their most complementary guests: spherands > cryptaspherands > cryptands > hemispherands > corands > podands. Spheraplex **137** $\cdot$ Li^+ provides a $-\Delta G^\circ$ value of >23 kcal mol^{-1}. The cryptaspheraplexes **144** $\cdot$ Na^+, **145** $\cdot$ Na^+, and **146** $\cdot$ Cs^+ (S. P. Ho[152]) provide values of 20.6, 21, and 21.7 kcal mol^{-1}, respectively.[146] Cryptaplexes **147** $\cdot$ Li^+, **148** $\cdot$ Na^+, and **135** $\cdot$ K^+ give values of 16.6, 17.7, and 18.0,[146] respectively. Hemispheraplexes **108** $\cdot$ Na^+, **149** $\cdot$ Na^+, and **150** $\cdot$ K^+ are bound by 12.2, 13.5, and 11.6 kcal mol^{-1}, respectively. Coraplex **151** $\cdot$ K^+ has a $-\Delta G^\circ$ value of 11.4, and podaplex **143** $\cdot$ M^+ values of <6 kcal mol^{-1}. Although the numbers of binding sites and their characters certainly influence these values, the degree of preorganization appears to be dominant in providing this order. Chart IV provides crystal structures of several cryptaspheraplexes.[152] (Compare **146** $\cdot$ Na^+ with **146** $\cdot$ K^+ $\cdot$ H_2O and **146** $\cdot$ Cs^+ $\cdot$ H_2O. In the former, the Na^+ does not contact all of the heteroatoms!)

Chart V contains the structures of a series of hosts containing mainly cyclic urea units as binding sites. In passing from **152** to **155** in studies with R. J. M. Nolte,[153] I. R. Dicker,[154] K. M. Doxsee, K. D. Stewart, M. Feigel, and J. W. Canary,[155] these units were found to be increasingly constrained to form a foundation for tripod binding of ammonium and alkylammonium guests. The ΔG°_{av} values for binding NH_4^+, $CH_3NH_3^+$, and $(CH_3)_3CNH_3^+$ picrates increased from 7.7 to 9.0 to 14.2 to 15.0 kcal mol^{-1}, respectively. Again, the effects of preorgani-

Chart III. Host structures, family names, and binding free energies of alkali metal picrate salts ($-\Delta G°$, kilocalories per mole, 25 °C, $CDCl_3$ saturated with D_2O)

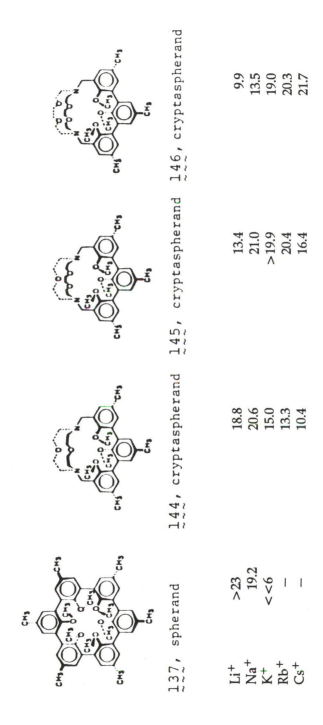

	137, spherand	144, cryptaspherand	145, cryptaspherand	146, cryptaspherand
Li+	>23	18.8	13.4	9.9
Na+	19.2	20.6	21.0	13.5
K+	<<6	15.0	>19.9	19.0
Rb+	—	13.3	20.4	20.3
Cs+	—	10.4	16.4	21.7

Continued on next page

Chart III. Host structures, family names, and binding free energies of alkali metal picrate salts ($-\Delta G°$, *kilocalories per mole, 25 °C, $CDCl_3$ saturated with D_2O)—Continued*

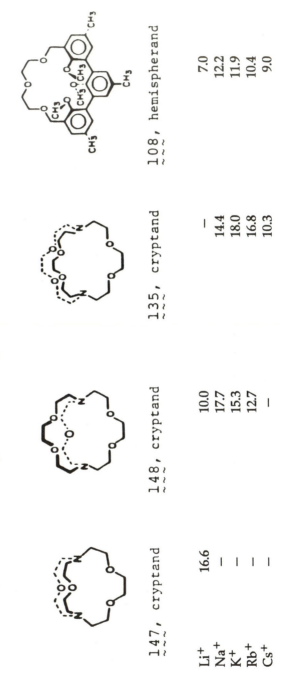

	147, cryptand	148, cryptand	135, cryptand	108, hemispherand
Li^+	16.6	10.0	—	7.0
Na^+	—	17.7	14.4	12.2
K^+	—	15.3	18.0	11.9
Rb^+	—	12.7	16.8	10.4
Cs^+	—	—	10.3	9.0

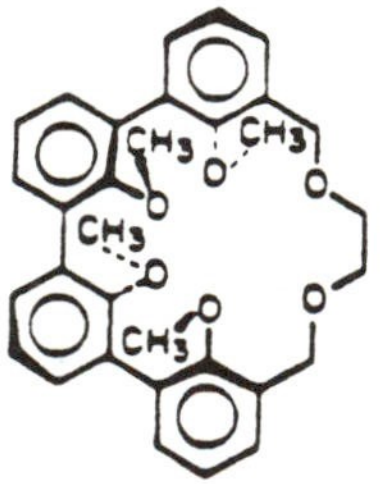

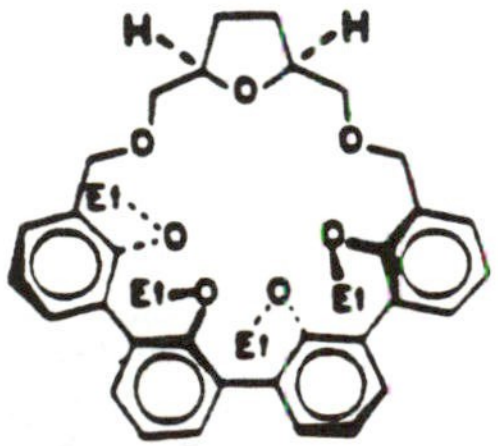

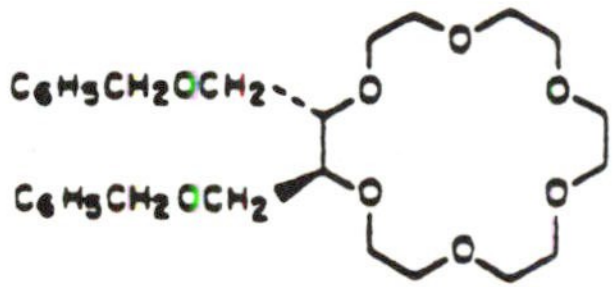

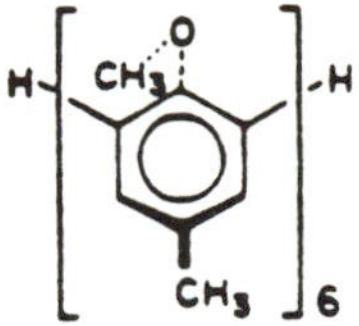

	149, hemispherand	150, hemispherand	151, corand	143, podand
Li$^+$	7.2	6.5	6.3	<6
Na$^+$	13.5	7.1	8.4	<6
K$^+$	10.7	11.6	11.4	<6
Rb$^+$	8.4	11.4	9.9	<6
Cs$^+$	7.1	10.8	8.5	<6

PROFILES, PATHWAYS, AND DREAMS

Chart IV. Stereorepresentations of crystal

144·Na⁺

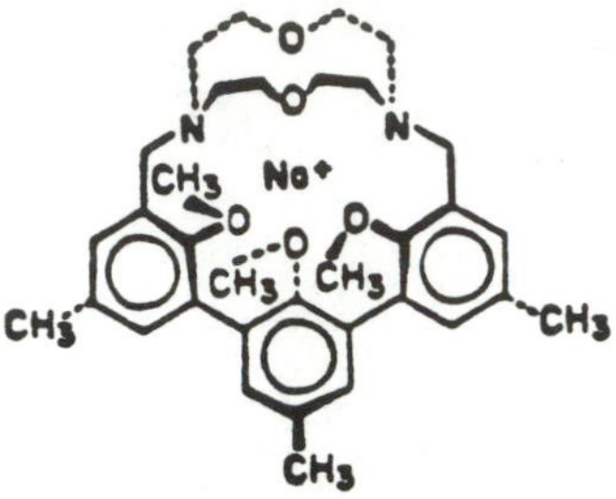

145·K⁺·H₂O

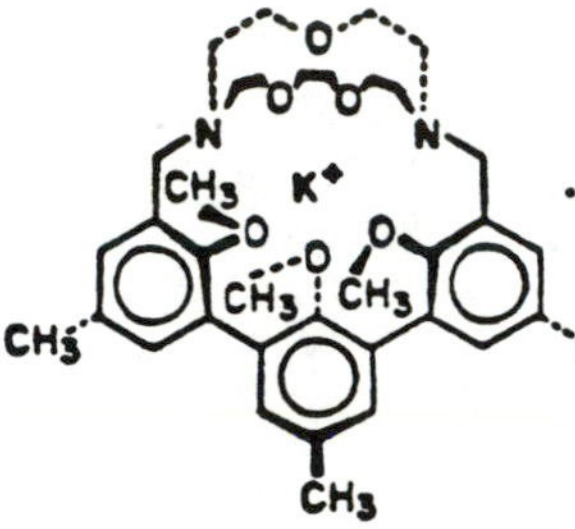

146·Na⁺

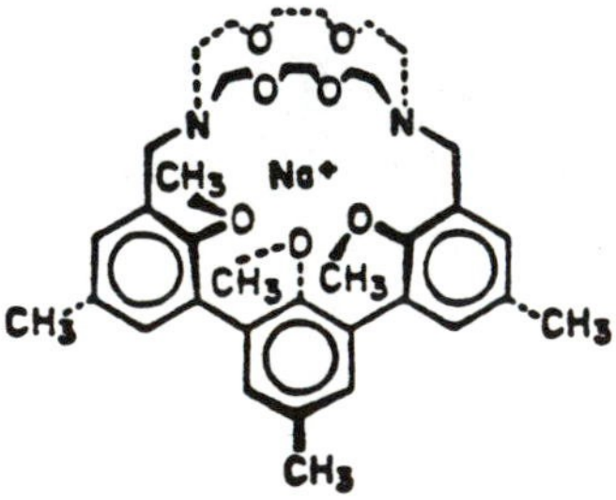

structures of cryptaspheraplexes

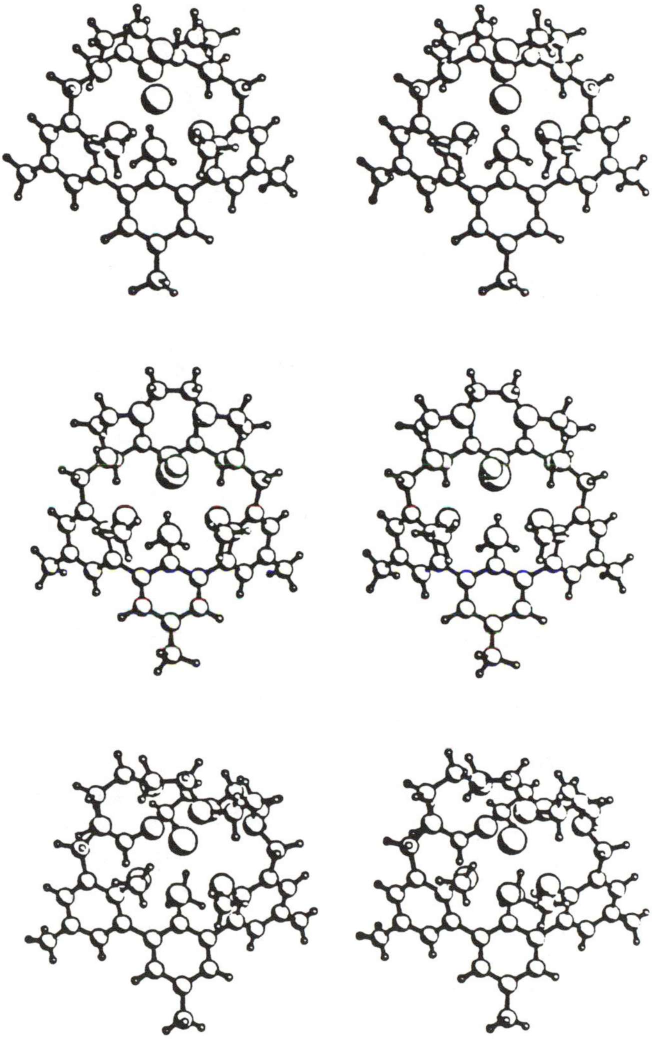

Continued on next page

Chart IV. Stereorepresentations of crystal

$\underset{\sim\sim\sim}{146} \cdot K^+ \cdot H_2O$ $\cdot H_2O$

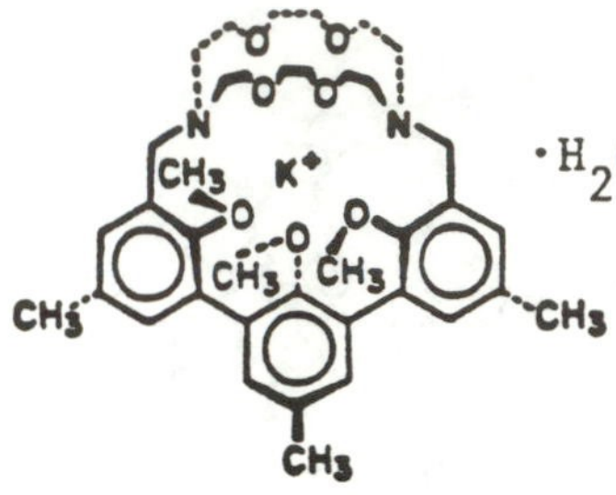

$\underset{\sim\sim\sim}{146} \cdot Cs^+ \cdot H_2O$ $\cdot H_2O$

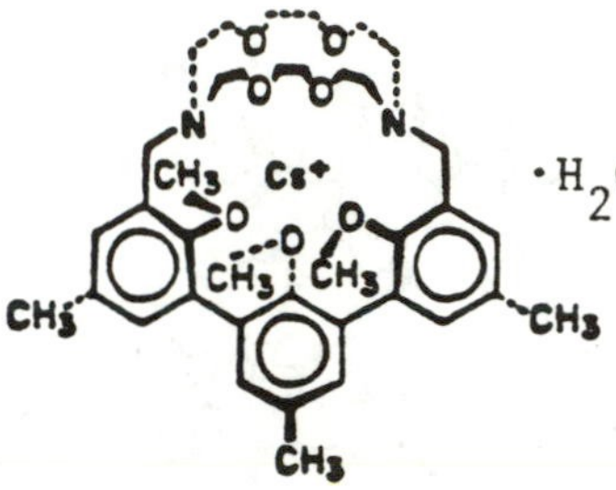

---

zation in forming perching complexes are evident. Host 152 has conformationally mobile binding sites subject to solvation from many directions, whereas 155 is rigid and can be solvated in only specialized ways.

The crystal structure of 156 (analogue of 152) is illustrated in drawing 157, and that of 158 (analogue of 155) is drawn in 159. The former host (156) lacks preorganization, whereas the latter (158) is nicely preorganized for binding.[156] The crystal structures of $154 \cdot (CH_3)_3CNH_3^+$ and $155 \cdot (CH_3)_3CNH_3^+$ beautifully illustrate the high levels of organization of the complexes.[154,155] The cyclic urea hosts are also strong binders of the alkali metal ions. Crystal structures of $160 \cdot Na^+ \cdot OH_2$, $160 \cdot Cs^+ \cdot OH_2$, and $161 \cdot Na^+ \cdot OH_2$ all contain 1 mol of water ligating the metal ion and hydrogen bonding nearby oxygens.[156] As predicted by CPK molecular model examination, complexes $154 \cdot (CH_3)_3CNH_3^+$, $155 \cdot (CH_3)_3CNH_3^+$, and $160 \cdot Cs^+ \cdot OH_2$ are all perching, whereas $160 \cdot Na^+ \cdot OH_2$ and $161 \cdot Na^+ \cdot OH_2$ are nesting (K. D. Stewart, M. Miesch).[157]

Our research struck a somewhat dissonant note in a paper published in 1981 entitled "Augmented and Diminished Spherands and Scales of Binding."[158] On the basis of CPK molecular model examination, we had expected the reactions of 162 and 165 to give 168 and 169,

structures of cryptaspheraplexes—Continued

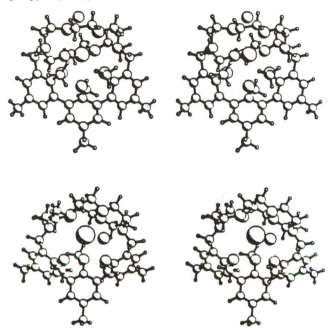

respectively. Models of **163** and **166** could not be assembled without grinding off about 15% of the four bridgehead oxygens, whereas those of **168** and **169** were easily assembled. Crystal structures **164** and **167** subsequently proved our initial structural assignments to be incorrect. Although very compressed and strained, spheraplexes **163** · Li$^+$ and **166** · Li$^+$ were the isomers isolated.[149] Although hosts **163** and **166** (like spherand **137**) formed capsular complexes only with Li$^+$ and Na$^+$, their $-\Delta G°$ values in our standard medium ranged only from 13.3 to 18.7 kcal mol^{-1}.[147] In **163** · Li$^+$, only five of the six oxygens contact Li$^+$, and in **166** · Li$^+$, only seven of the eight oxygens contact Li$^+$. Thus, CPK molecular models proved misleading in predicting the structures of compounds prepared from high-energy reaction intermediates. In the reaction we invented to prepare these spherands,[149] aryl–lithium reagents are oxidized with ferric acetylacetonide to form what are probably aryl radicals that couple to give spheraplexes.

Molecular mechanical calculations on **168** · Li$^+$ confirm that it is much more stable than its *syn* isomer, **163** · Li$^+$, and even more stable than **137** · Li$^+$. We have since prepared **168** and **169** and have demonstrated that **168** · Li$^+$ is more stable (as predicted)[159] than **137** · Li$^+$.[160] Rate constants k_1 and k_{-1} for complexation–decomplexation (equation

Chart V. Binding free energies ($-\Delta G°$, kilocalories per mole, 25 °C, $CDCl_3$ saturated with D_2O) of hosts containing cyclic urea units

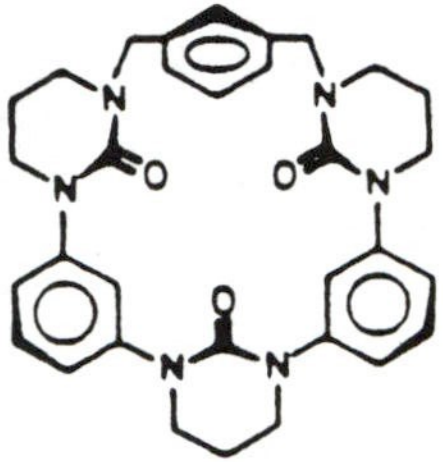

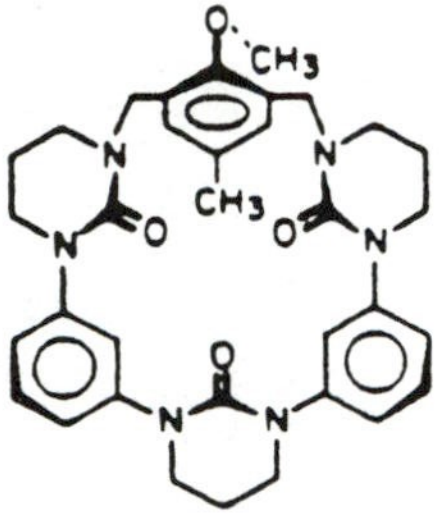

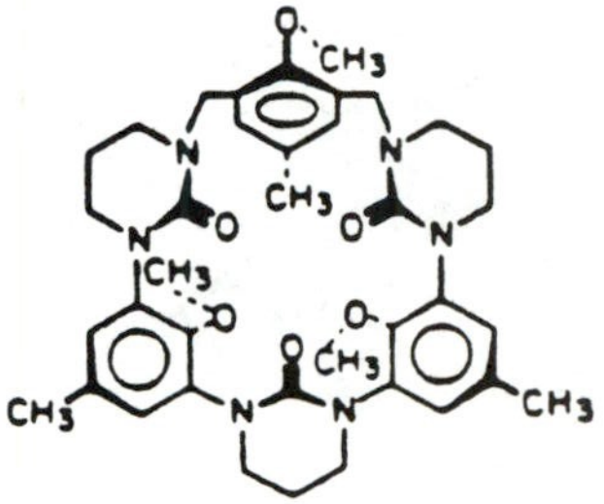

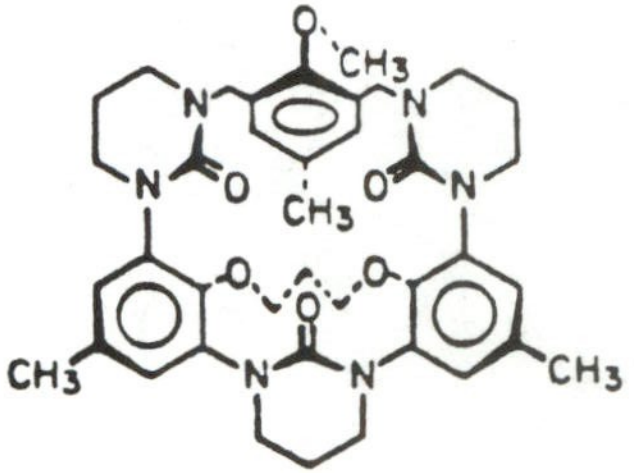

	152	**153**	**154**	**155**
NH_4^+	7.0	8.4	14.4	15.7
$CH_3NH_3^+$	7.4	9.0	14.4	14.9
$(CH_3)_3CNH_3^+$	8.6	9.5	13.2	14.3
$\Delta G°$ av.	7.7	9.0	14.2	15.0

10) were determined at 25 °C in $CDCl_3$ saturated with D_2O for enough host–guest pairs (S. P. Artz,[161] T. Anthonsen,[162] and G. M. Lein[147]) to provide several generalizations:

1. The more highly preorganized the hosts are for complexation, the lower are the rate constants for complexation–decomplexation.

2. Rate constant values for either formation or dissociation of complexes, when arranged in decreasing order, provide the sequence perching > nesting > capsular.

3. All of the complexation processes studied in homogeneous solution are instantaneous on the human and ^1H NMR time scales.

The decomplexation rates of the hemispheraplexes and coraplexes are fast on the human and slow on the ^1H NMR time scales. The decomplexation rates of the spheraplexes and the more stable cryptaspheraplexes are very slow on the human time scale at 25 °C.

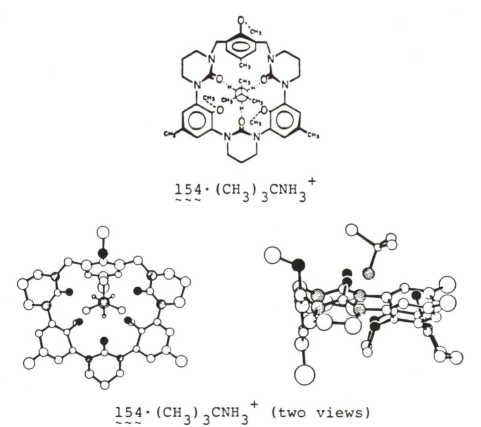

$$\underset{\sim\sim\sim}{154} \cdot (CH_3)_3CNH_3^+$$

$$\underset{\sim\sim\sim}{154} \cdot (CH_3)_3CNH_3^+ \quad \text{(two views)}$$

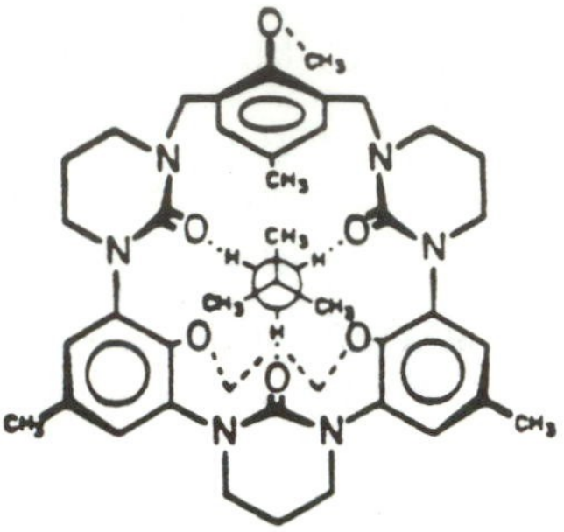

$\underset{\sim\sim\sim}{155} \cdot (CH_3)_3CNH_3^+$

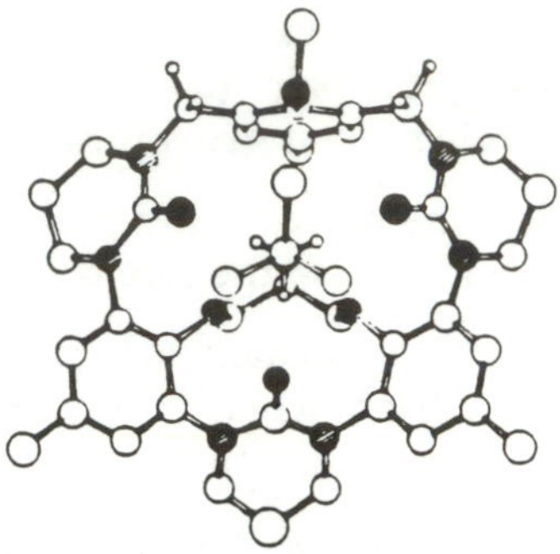

$\underset{\sim\sim\sim}{155} \cdot (CH_3)_3CNH_3^+$

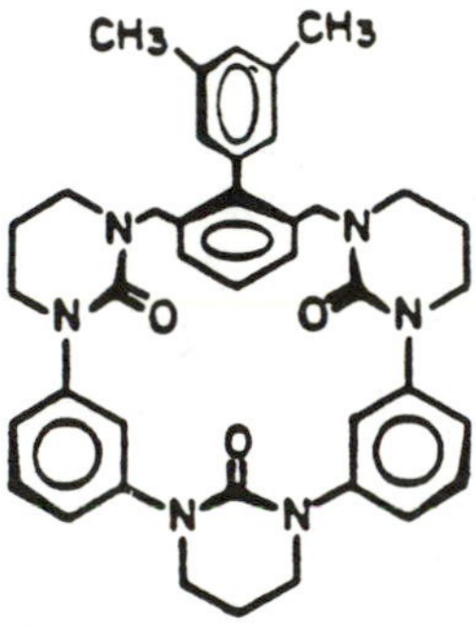

$\underset{\sim\sim\sim}{156}$

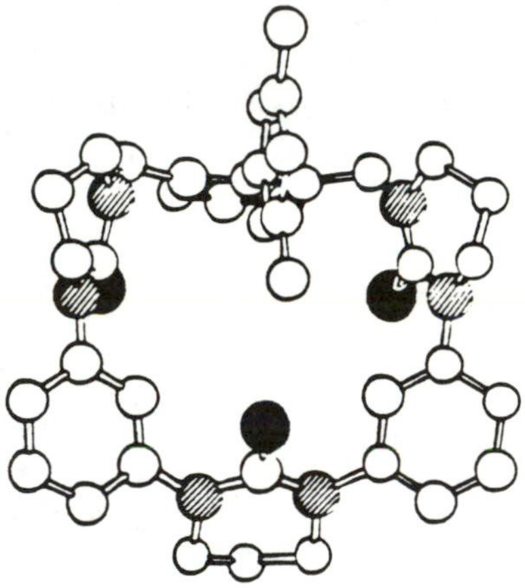

$\underset{\sim\sim\sim}{157}$

I thoroughly enjoyed supervising the thesis work of G. M. Lein, S. P. Artz, K. D. Stewart, H. E. Katz, S. P. Ho, and I. R. Dicker in the studies discussed in this section, and the colleagueship of T. Kaneda, R. C. Helgeson, S. B. Brown, T. Anthonsen, R. J. M. Nolte, K. M. Doxsee, C. B. Knobler, E. F. Maverick, I. Goldberg, and K. N. Trueblood. The crystal structures determined by the last four people still inspire in me a sense of magic of first seeing the long unseen.

Structural Recognition and the Principle of Complementarity

Just as preorganization is the central determinant of binding power, complementarity is the central determinant of structural recognition.

The binding energy at a single contact site is, at most, a few kilocalories per mole, much lower than that of a covalent bond. Contacts at several sites between hosts and guests are required for the structuring of complexes. Such contacts depend on complementary placements of binding sites in the complexing partners. The self-evident principle of complementarity has been stated as follows: "To complex, hosts must have binding sites which cooperatively contact and attract binding sites of guests without generating strong nonbonded repulsions."[147]

The most extensive correlations of structural recognition with host–guest structure involve the K_a values with which the spherands, cryptaspherands, cryptands, hemispherands, corands, and podands associate with the various alkali metal picrate salts at 25 °C in $CDCl_3$ saturated with D_2O. The constants for distribution of the five alkali metal ion picrate salts between D_2O and $CDCl_3$ in the absence of host differ from one another by factors of less than 2 for alkali metal cations that neighbor one another in the periodic table. The largest factor involving any two of the five ions is only 3.8.[120] Thus, $K_a^A/K_a^{A'}$ ratios for two alkali metal ions A and A' in $CDCl_3$ saturated with D_2O provide a rough measure of ion selectivity in extraction, as well as in homogeneous media. Chart VI summarizes the largest $K_a^A/K_a^{A'}$ ratios that we have obtained for neighboring alkali metal ions.[163] (None of the hosts listed are either corands or podands.)

Arrangement of the classes of hosts in decreasing order of their ability to select between the alkali metal ion guests provides spherands

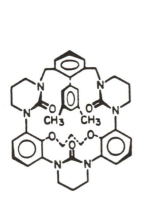

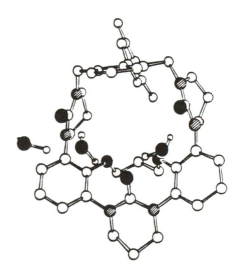

158   159

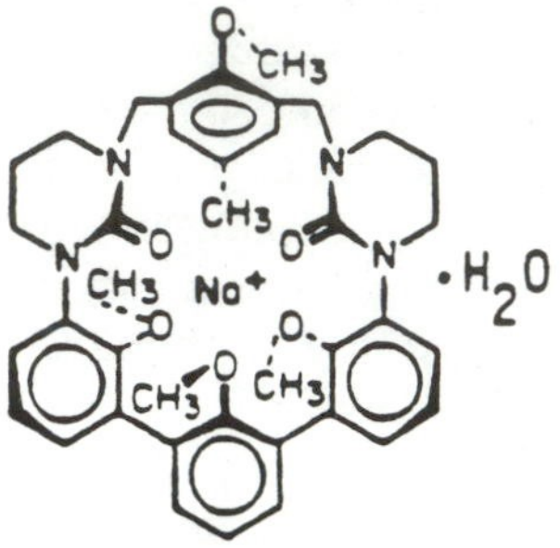

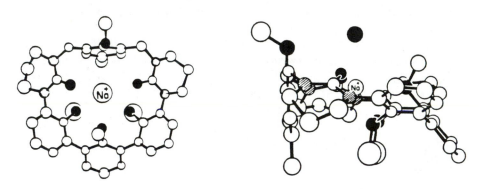

160·Na$^+$·OH$_2$

160·Na$^+$·OH$_2$

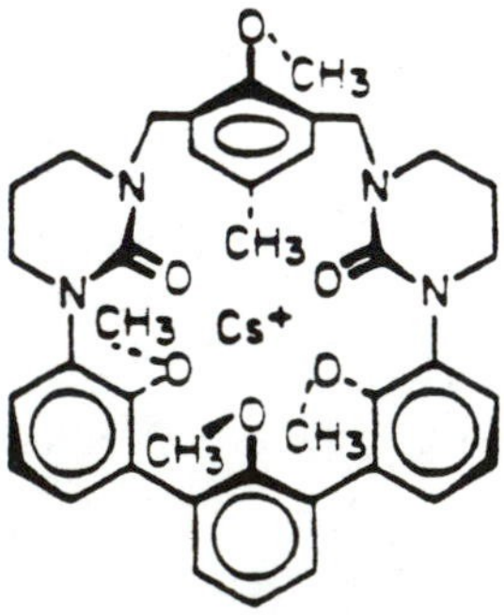

160·Cs$^+$·OH$_2$

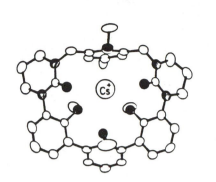
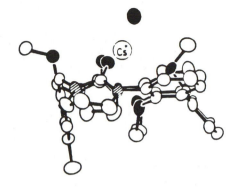

$$\underset{\sim\sim\sim}{160}\cdot Cs^{+}\cdot OH_2$$

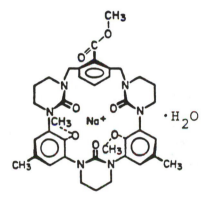

$$\underset{\sim\sim\sim}{161}\cdot Na^{+}\cdot OH_2$$

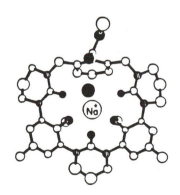
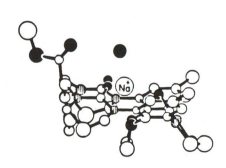

$$\underset{\sim\sim\sim}{161}\cdot Na^{+}\cdot OH_2$$

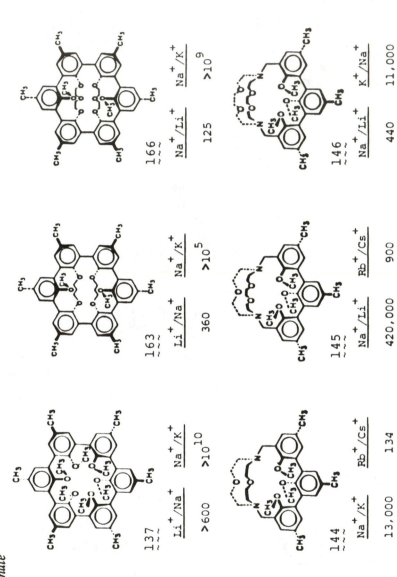

Chart VI. Structural recognition measured by $K_a^A/K_a^{A'}$ values for alkali metal picrates at 25 °C in D_2O saturated with D_2O. Starred values are crude estimates based on analogies between our standard medium and propylene carbonate

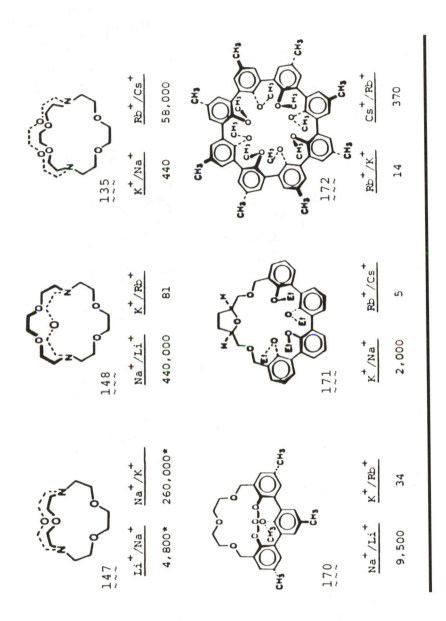

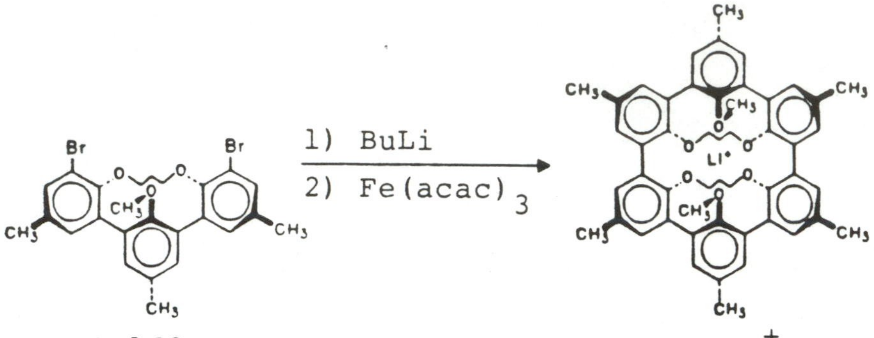

162

163·Li$^+$

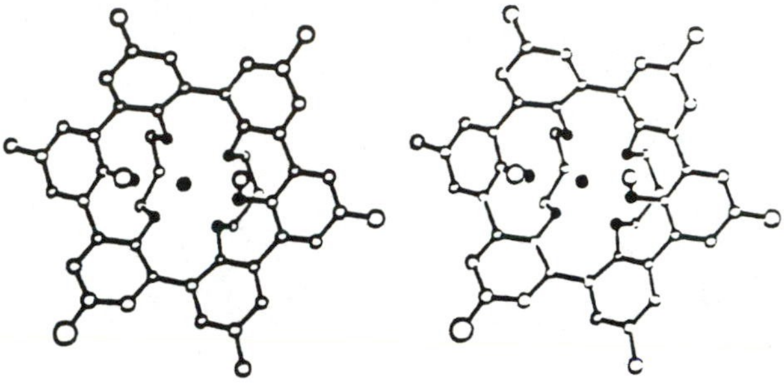

164

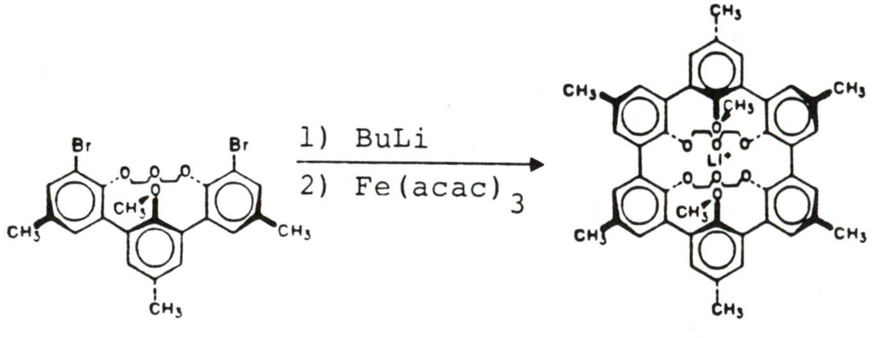

165

166·Li$^+$

> cryptaspherands ~ cryptands > hemispherands > corands > podands. This order is similar, but less rigidly followed, than that for host preorganization. In some cases, rather small changes in structure provide a substantial spread in $-\Delta G°$ values for binding under our standard conditions. Examples are found in the patterns of $-\Delta G°$ value–structural relationship found in Figures 2–4.

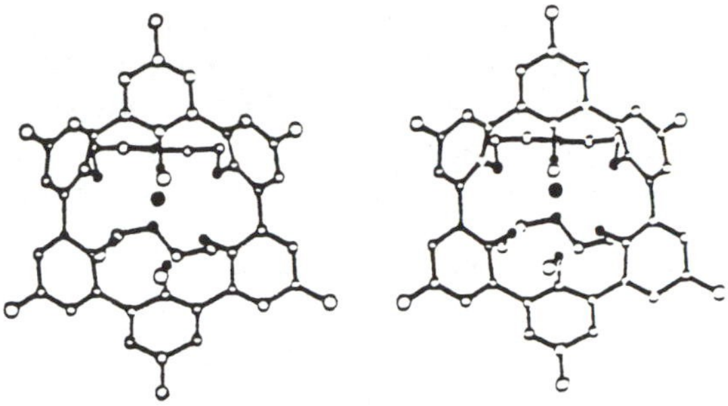

167

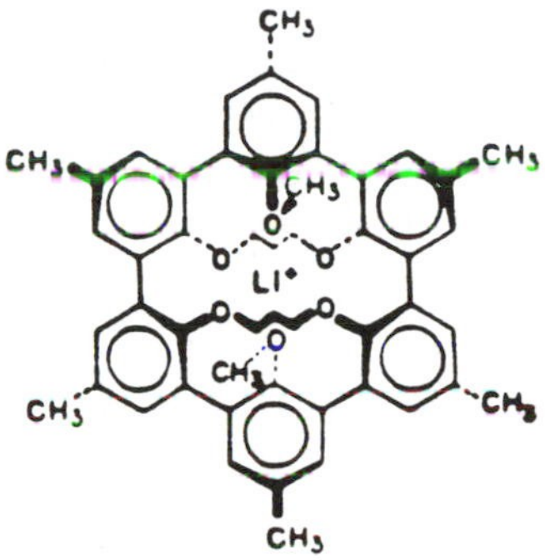

168·Li⁺

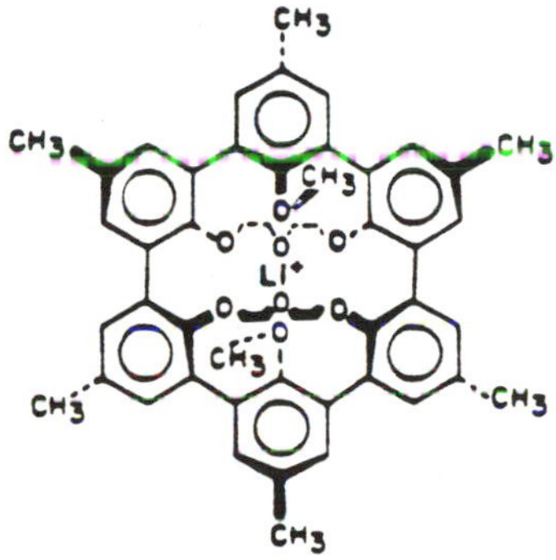

169·Li⁺

All of the hosts of Figure 2 possess 18-membered macrorings.[164] As anisyl oxygens are successively substituted for corand binding sites (e.g., **174**) the hosts become less selective and weaker binders (e.g., **173**) until three contiguous anisyls are present (as in **108**), at which point the hosts become stronger and more selective complexing agents. They pass from the corand to the hemispherand class. At this point, the anisyls start to become self-organizing for steric reasons. Addition of an extra bridge to the hemispherand (as in **170**) increases the binding, but not the selectivity. Hemispherand **149**, containing four anisyl groups preorganized, shows strong and selective binding toward Na^+, to which it is complementary (S. P. Artz).[161] Figure 3 traces changes of ion selectivity and binding power as cyclic urea and anisyl molecular modules are juggled, and macroring sizes are changed (I. B. Dicker and M. Lauer).[154] Host **154** contains a 20-membered macroring and, in molecular models, appears adaptable to guest character. This hemispherand binds all eight guests strongly with little structural recognition. The binding sites of

160 differ from 154 only in the sense that an anisyl unit at 6 o'clock in 160 has been substituted for a cyclic urea unit in 154. As a result, 160 shows more selectivity but weaker binding. Contiguous anisyl groups are more self-organizing than contiguous cyclic urea–anisyl groups, although cyclic urea groups are intrinsically better ligating groups than

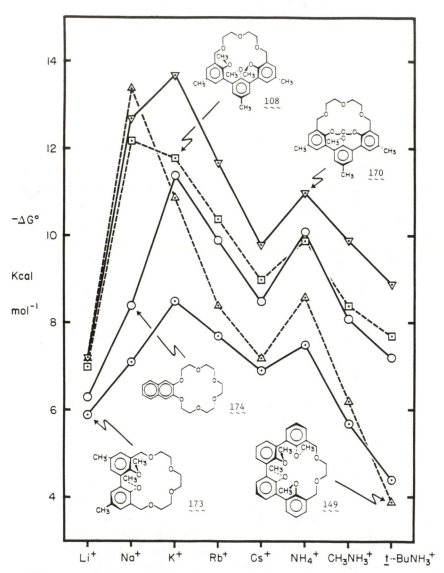

Figure 2. Changes in patterns of ion selectivities and binding power as host structures are changed from corand to hemispherand (25 °C in DCCl$_3$ saturated with D$_2$O).

are anisyls. Host **171** shows very high specificity for binding K$^+$ over Na$^+$ and low specificity for Rb$^+$ over Cs$^+$. Notice that **172** of Chart VI contains a 24-membered ring and shows high specificity in binding Cs$^+$ over binding Rb$^+$. When the macroring sizes are decreased to 19 members, (as in **175** and **176**) selectivity for binding the small ions increases markedly. Corand **174** is a weaker binder than any of the

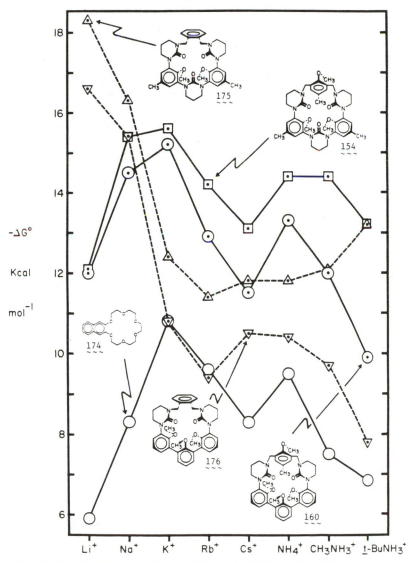

Figure 3. Changes in patterns of ion selectivity and binding power as host structures are changed (25 °C in DCCl$_3$ saturated with D$_2$O).

hemispherands of Figure 3 and exhibits less selectivity than the more highly preorganized systems **175** and **176**.

The relationships of Figure 4 illustrate several points.[154] Host **177** contains good binding modules, but the locations of the two methylene groups provide the system with too much conformational mobility for it to be a stronger or more selective binder than standard lipophilic corand **174**. Although **178** has a similar location of methylenes, the spatial requirements for its two benzyl groups provide the system with a self-organizing feature evident in its becoming a stronger and more selective complexing compound. The internal hydrogen bond of the phenolic group in **180** reduces its binding power and selectivity as compared to **179**, whose space-occupying benzyl and allyl groups provide some self-organization to the system.[154] A large number of hosts had to be designed and synthesized and had their binding abilities compared under standard conditions before these structure–binding correlations emerged. I designed many hosts that my co-workers either failed to synthesize, or did not try to prepare because of the absence of rational syntheses. Many complexes that we did prepare failed to provide crystals appropriate for crystal structure determination. What emerged was the possible, a blend of what we did and did not anticipate.

The question of scientific yield versus effort expended is wholesome and constantly occurring. I always pose this question when I read theses, write grant applications, write review articles, and particularly while composing this volume. The principles of preorganization and complementarity are self-evident and very simple concepts; however, they gain credibility only through repeated exemplification and testing in a variety of contexts. Until and unless a large body of coherent structure–behavior data is in-hand, the concepts are empty. We expended great effort in generating this data, which emerged piecemeal. The scientific yield per unit effort appeared low for some of the pieces, but when merged with the other pieces, the patterns appeared and the yield was gratifying. I greatly enjoyed this research, and am happy with the results.

Partial Transacylase Mimics

The design and synthesis of enzyme-mimicking host compounds remains one of the most challenging and stimulating problems of organic chemistry. I chose to examine transacylase mimics first because the mechanism of action of these enzymes had been so thoroughly studied. The first system was examined (1976) by Y. Chao,[165,166] G. R. Weisman,[166] and G. D. Y. Sogah[166] and involved **181**, which is chiral,

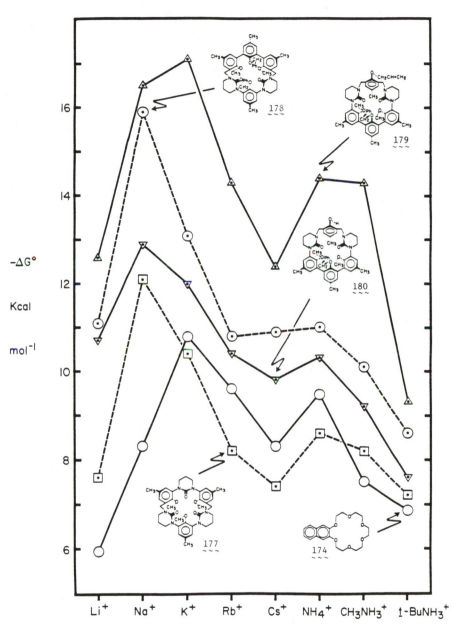

Figure 4. *Changes in patterns of ion selectivities and binding power as host structures become more self-organizing (25 °C in DCCl₃ saturated with D₂O).*

contains a C_2 axis to make it nonsided, and combines a good binding site for an amine group (like trypsin) with a good nucleophile (like papain). The crystal structure of 106[119] shows that the methyl group of the host is in between the two methyl groups of the guest. Model examination suggested that (L)-amino esters such as 182 should form complexes resembling 106 in which the CH_2SH group is proximate to the carbonyl group of the guest. Thus, complexation turns two remote functional groups into neighboring groups, a process we call collecting and orienting. The poorly binding podand 183 was prepared for comparison purposes.

The acylation of 181 by 182 was found to occur about 1000 times faster than acylation of 183. Unlike 183, corand 181 showed chiral recognition by catalyzing the reaction of the (L)-enantiomer of 182 more than that of the (D)-enantiomer by a factor of about 10. The reaction to form 184 was subject to competitive inhibition by added K^+, with which 181 complexes. Reaction of 184 with EtOH to give 185 was much slower.[165,166] Somewhat later, K. Koga et al.[167] and J.-M Lehn et al.[168] reported similar results with other corand hosts with attached thiol groups.

The active site of chymotrypsin combines a binding site, a nucleophilic hydroxyl, an imidazole, and a carboxyl group in an array preorganized largely by hydrogen bonds (as indicated in 186). With the help of molecular models, I designed 187 as an "ultimate target" host, possessing roughly the same organization of groups as that of 186.

Compound 187 is much too complicated to synthesize without getting encouragement from simpler model compounds. An incremental approach to 187 was employed. I. B. Dicker and I[154] first prepared 154 and found that it binds t-$C_4H_9NH_3$Pic in $CDCl_3$ saturated with D_2O with $-\Delta G\,^{\circ} = 13.2$ kcal mol^{-1}. The complex, $154\cdot(CH_3)_3CNH_3{}^+$, had the expected crystal structure. Accordingly, 188 was prepared by H. E. Katz,[169] and found to bind CH_3NH_3Pic and NaPic (under our standard conditions) with $-\Delta G\,^{\circ} = 12.7$ and 13.6 kcal mol^{-1}, respectively. Host 188 was acylated by 189 to give 190 and p-nitrophenol. The kinetics of formation of 190 were measured in $CHCl_3$ and found to be first-order in added Et_3N–Et_3NHClO_4 buffer ratio. Thus, the alkoxide ion was the nucleophile. The rate constant for acylation of 188 by 189 was calculated to be $\sim 10^{11}$ higher valued than the rate constant for the noncomplexed model compound, 3-phenylbenzyl alcohol.[170] This high factor demonstrates that collecting and orienting reactants through highly structured complexation can result in an enormous rate acceleration. When $NaClO_4$ was added to the medium, the acylation rate of 188 was depressed by several powers of 10; thus, the acylation of 188, like that of the serine esterases, is subject to competitive inhibition.

A 30-step synthesis of 191 was then devised, and about 0.5 g of

106

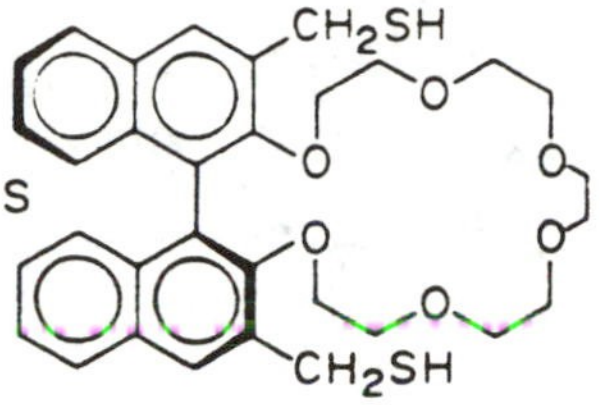

181 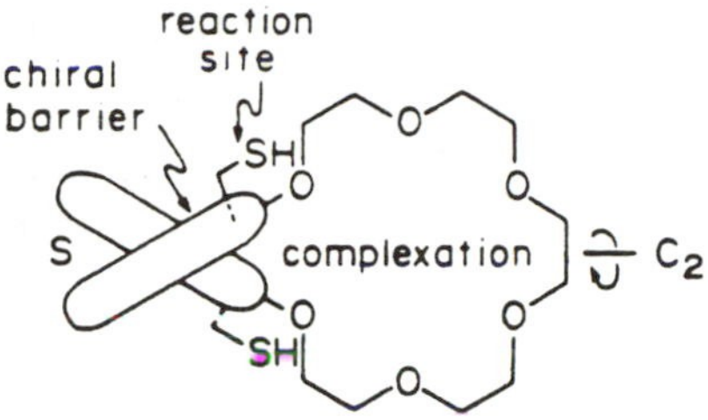   181

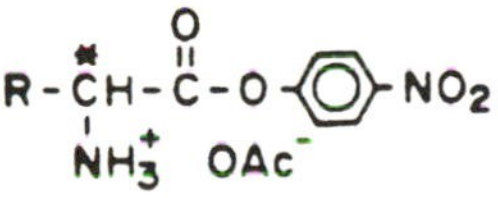

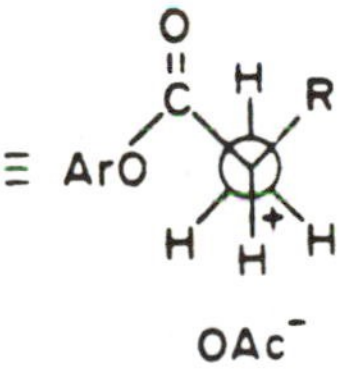

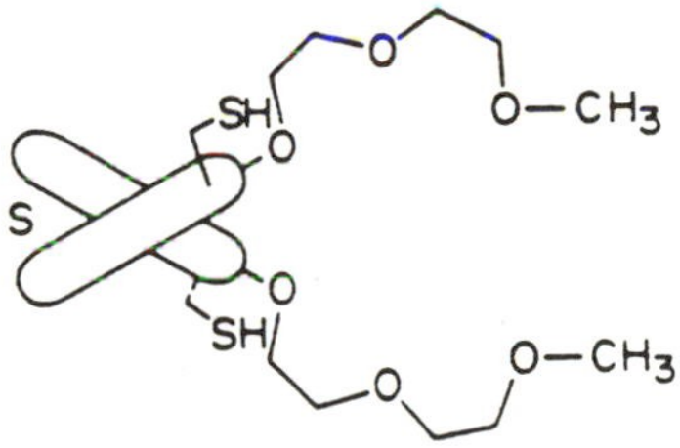

L-182   L-182   183

181 + 182

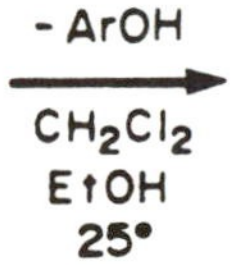

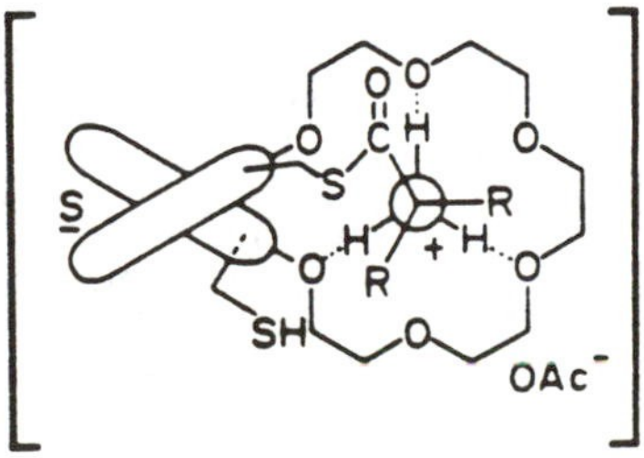

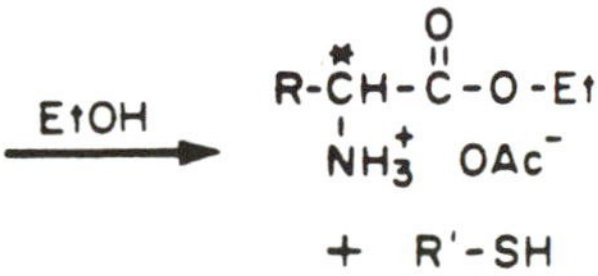

$\xrightarrow[\substack{CH_2Cl_2 \\ EtOH \\ 25°}]{- ArOH}$

$\xrightarrow{EtOH}$

R-CH-C-O-Et
  NH₃⁺ OAc⁻

+ R'-SH

184   185

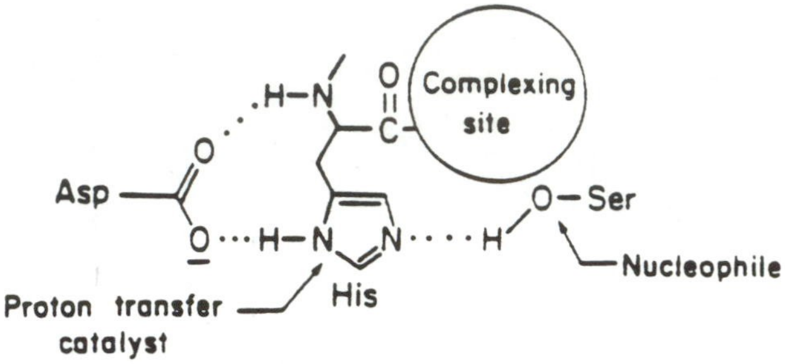

186

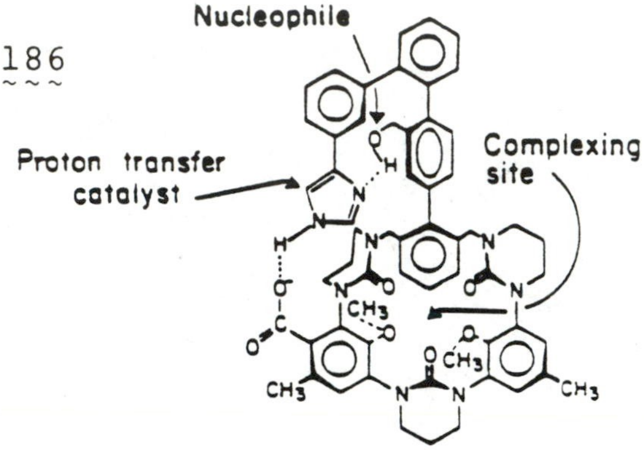

187

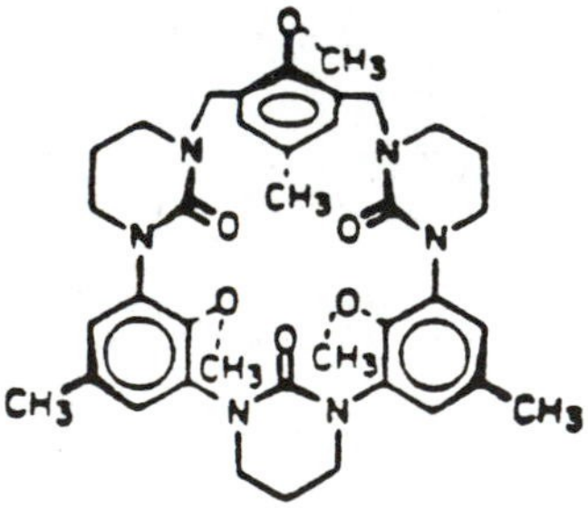

154

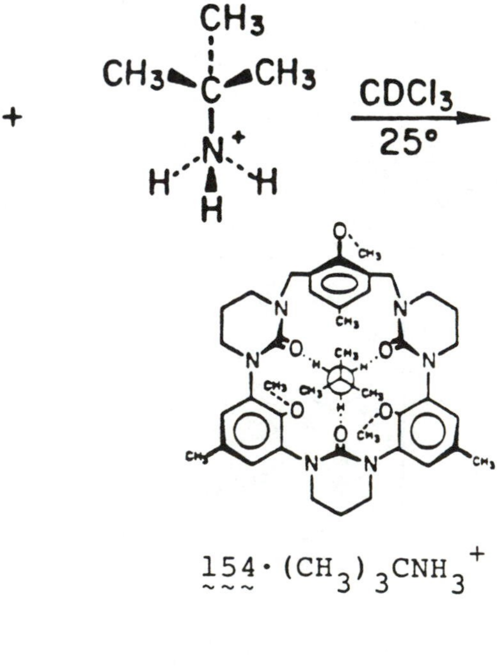

154 · (CH$_3$)$_3$CNH$_3$$^+$

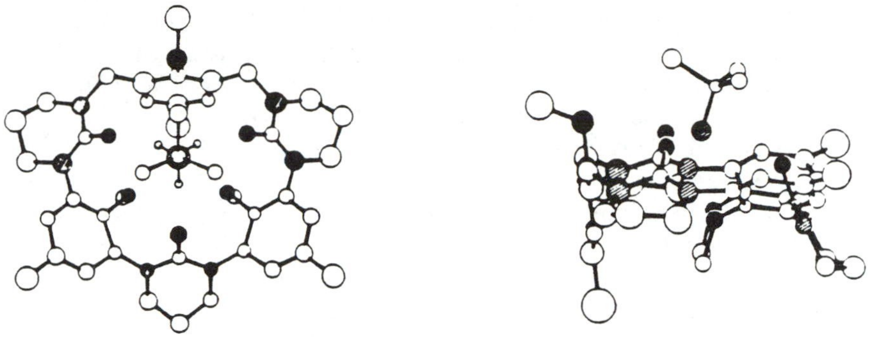

$$\underset{\sim\sim\sim}{154} \cdot (CH_3)_3CNH_3{}^+$$

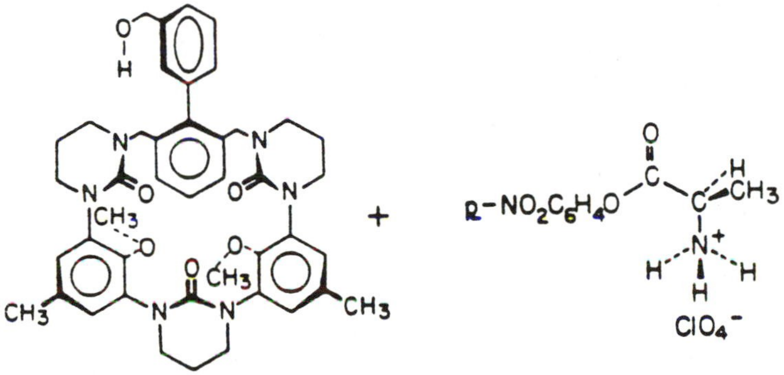

$$\underset{\sim\sim\sim}{188}$$

+

$$\underset{\sim\sim\sim}{189}$$

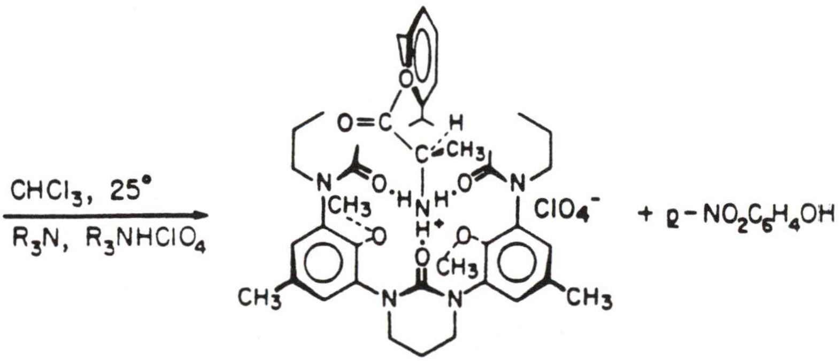

$$\xrightarrow[\text{R}_3\text{N, R}_3\text{NHClO}_4]{\text{CHCl}_3,\ 25°}$$

$$+\ \varrho-NO_2C_6H_4OH$$

$$\underset{\sim\sim\sim}{190}$$

the compound prepared by P. Y. S. Lam and I.[171] This compound combined the binding site, the nucleophilic hydroxyl, and the imidazole proton-transfer agent in the same molecule, and lacked only the carboxyl group of final target compound **187**. Compound **191** complexed CH_3NH_3Pic and NaPic with $-\Delta G°$ values of 11.4 and 13.6 kcal mol^{-1} in $CDCl_3$ saturated with D_2O at 25 °C. In pyridine–chloroform, amino ester salt **192** instantaneously acylated the imidazole group of **191** to give **193**, which more slowly gave **194**. In CHCl$_3$, in the absence of any added base, the observed rate constant for acylation of **191** by **192** was higher by a factor of 10^5 than that for acylation of an equal molar mixture of noncomplexing model compounds **197** or **198** under the same conditions. The same ratio was obtained when **195** was substituted for **191**; thus, the imidazole groups of **191** and **195** are the sites of acylation. Introduction of NaClO$_4$ into the medium as a competitive inhibitor of complexation destroyed much of the rate acceleration. When **192** added to **196** was substituted for **191**, the resulting complex acylated imidazole **198** with a 10^3 rate-constant factor increase. Thus, complexed **192** is a better acylating agent than **192** alone.

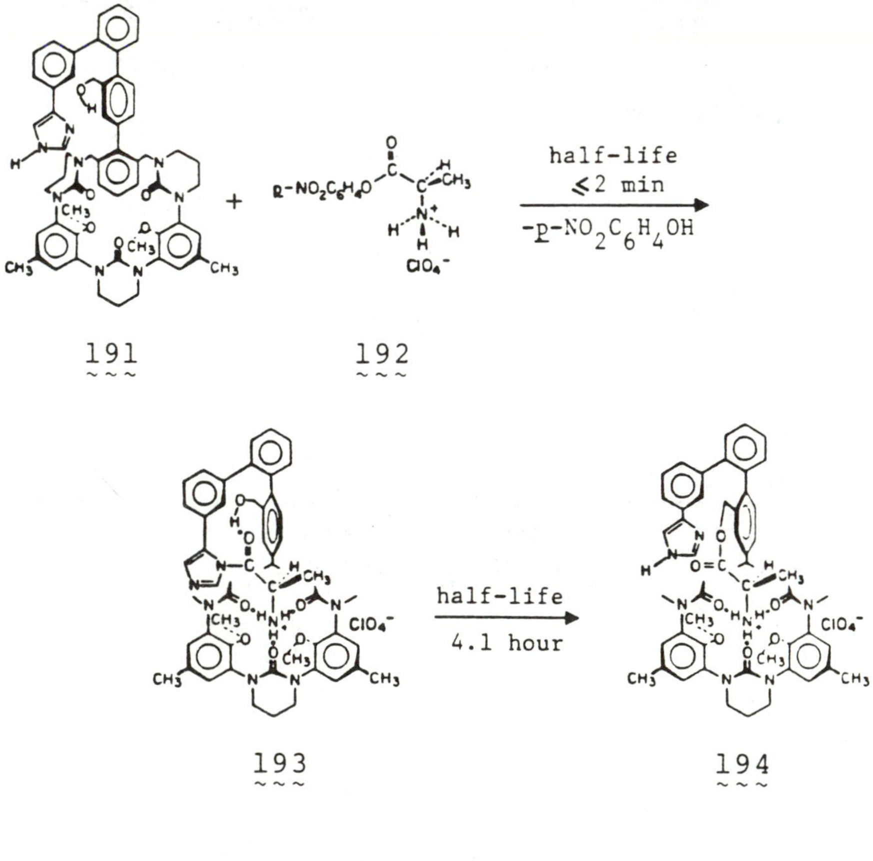

$$191 \qquad\qquad 192$$

$$193 \qquad\qquad\qquad 194$$

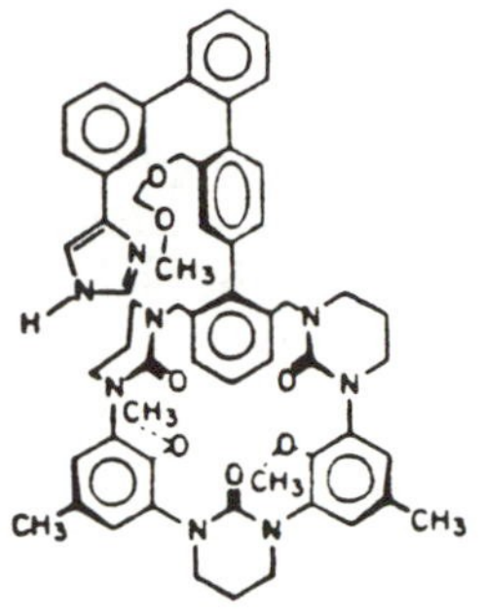

195

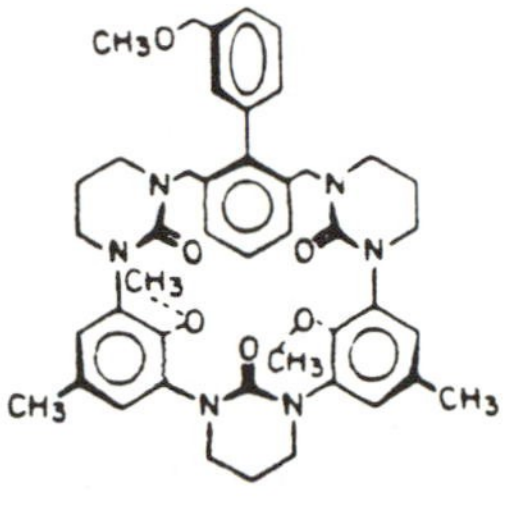

196

197 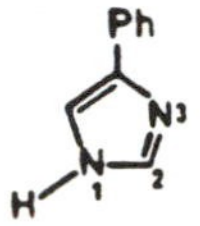   198

The disadvantages of comparing rate constants for reactions with different molecularities are avoided by referring to uncomplexed **191** or **195**, noncomplexing imidazole **198**, and uncomplexed acylating agent **192** as standard starting states and the rate-limiting transition states for transacylation as standard final states. This treatment introduces K_a into the second-order rate-constant expression when complexation precedes acylation. The resulting second-order rate constants for **192** acylating **191** or **195** are higher by factors of 10^{10} or 10^{11} than the second-order rate constant for **192** acylating **198**. This work clearly demonstrates that complexation of the transition states for transacylation can greatly stabilize those transition states to produce large rate factor increases over comparable noncomplexed transition states.[172] Others have shown that the imidazole of chymotrypsin is acylated first by esters of nonspecific substrates.[173]

These investigations demonstrate that totally synthetic systems can be designed and prepared that mimic the following properties of enzymes: the ability to use complexation to vastly enhance reaction rates, the capacity to distinguish between enantiomeric reactants, and the sensitivity to competitive inhibition. We anticipate that as the field matures, many of the other remarkable properties of enzyme systems will be observed in designed, synthetic systems. Our results illustrate some of the strategies and methods that might be applied in this expanding field of research.

Cavitands—Synthetic Molecular Vessels

Although enforced cavities of molecular dimensions are frequently encountered in enzyme systems and RNA or DNA, they are almost unknown among the 6 million synthetic organic compounds. In biological chemistry, such cavities play the important role of providing concave surfaces to which are attached convergent functional groups that bind substrates and catalyze their reactions. If synthetic biomimetic systems are to be designed and investigated, simple means must be found of synthesizing compounds containing enforced concave surfaces of dimensions large enough to embrace simple molecules or ions. I applied the name *cavitand* to this class of compound,[174] after the term had passed "Cram's test."

Cavitands designed and studied in my research group included compounds **200–204**, many of which were prepared from **199**. The structure and conformational mobility of **199** had been established by A. G. S. Högberg[175] for his Ph.D. thesis before he carried out research in my group on azacorands. The substance was prepared in good yield by treatment of resorcinol with acetaldehyde and acid. We rigidified **199** and its derivatives by closing four additional rings to produce **200–203** (S. Karbach and J. R. Moran,[176] and K. D. Stewart et al.[177]).

As anticipated by molecular model examinations, **200–203** crystallize only as solvates because these rigid molecules taken alone are incapable of filling their voids either intermolecularly or intramolecularly. They are shaped like bowls of differing depth supported on four methyl "feet." Compound **200** forms crystallites with SO_2, CH_3CN, and CH_2Cl_2 molecules to which it is complementary (molecular model examination). Cavitand **201**, whose cavity is deeper, crystallizes with 1 mol of $CHCl_3$. Crystal structures of $200 \cdot CH_2Cl_2$ and $201 \cdot CHCl_3$ show they are caviplexes[178] as predicted. Cavitand **202** is vase-shaped. It crystallizes with 1 mol of $(CH_3)_2NCHO$, which is just small enough to fit into the interior of **202** in models. Although the amide cannot be removed at high vacuum and low pressure, it is easily displaced with $CHCl_3$, 1.5 mol of which appear to take the place of the $(CH_3)_2NCHO$ in the crystallite.[176]

Treatment of octaphenol **199** with R_2SiCl_2 gave a series of cavitands, of which **203** is typical. In molecular models, **203** has a well-shaped cavity, defined by the bottoms of four aryls and by four inward-turned methyl groups. In molecular models, this well is complementary to small cylindrical molecules such as $S=C=S$, $CH_3C\equiv CH$, and $O=O$ but not to larger compounds such as $CDCl_3$ or C_6D_6. Cavitand **203** and its analogues, when dissolved in $CDCl_3$ or C_6D_6, complex guests such as those just mentioned, whose external surfaces are complementary to the internal surface of the host cavity. Association con-

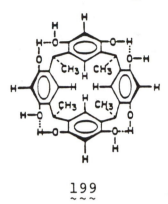

199

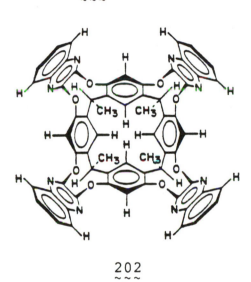 200

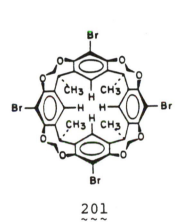

201

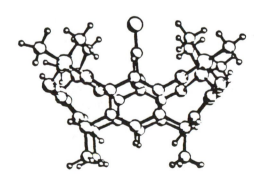 202

203

$203 \cdot CS_2$ (side view)

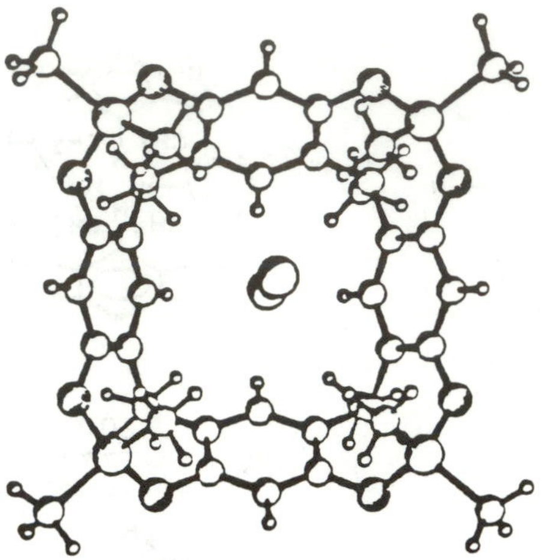

$\underset{\sim\sim\sim}{203} \cdot CS_2$ (top view)

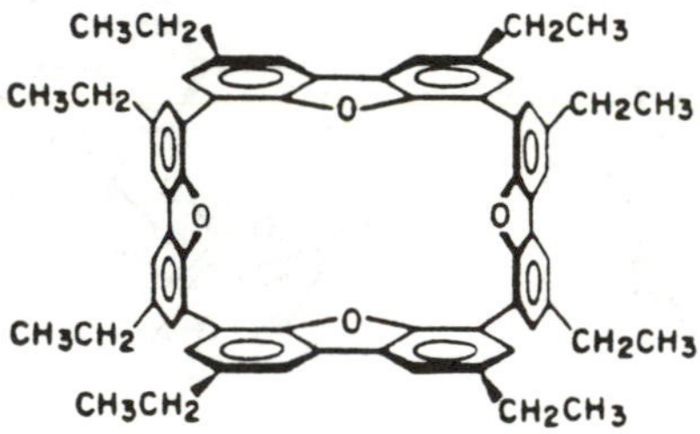

$\underset{\sim\sim\sim}{204}$

stants were determined for **203** and its analogues for binding S=C=S. Values of $-\Delta G°$ as high as 2 kcal mol^{-1} have been observed. A crystal structure of **203** · CS$_2$ shows that CS$_2$ occupies the well in the expected manner. Compound **203** in CDCl$_3$ was also shown to bind dioxygen reversibly.[177] Dissolution of **203** in solvents such as CDCl$_3$ or C$_6$D$_6$ is the equivalent of dissolving "holes" in a medium into which appropriately shaped solutes fall. The discrimination shown by the holes for the guests exemplifies the principle of complementarity as applied to complexation.

Cavitands **204** and **205** were also synthesized by very simple reactions (R. C. Helgeson, M. Lauer, and I).[179] A molecular model of **204** shows that it contains two long, cleft-shaped cavities approximately 10.8 Å long × 3.4 Å wide × 4.3 Å, deep, each being large enough to easily contain biphenyl. A model of **205** possesses a cavity large enough

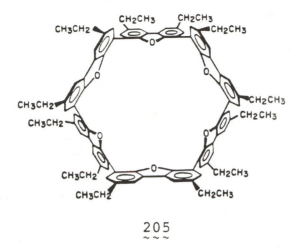

$$205$$

to embrace seven models of benzene, one coplanar with and complementary to its six oxygens, three benzenes stacked on one side, and
three on the other side of the central benzene. Experimentally, both **204**
and **205** are easily manipulated, and form crystallites with benzene and
other aromatic compounds. Interestingly, cavitands **200–202** are complementary to CH_2Cl_2 and $CHCl_3$ and are soluble in these solvents, but
are insoluble in C_6H_6, to which they are noncomplementary. Con-
versely, **204** and **205** are complementary to C_6H_6 and $C_6H_5CH_3$ and soluble in these solvents, but are insoluble in CH_2Cl_2 and $CHCl_3$, to which
they are noncomplementary.[177,180]

Cavitands **206** and **207** are chiral (R. C. Helgeson, J.-P. Mazaleyrat,
and I).[180] In molecular models, their methyl groups can turn either
inward or outward. Molecular models of **206** with its methyl groups
oriented outward is complementary to chair cyclohexane and when
oriented inward is complementary to diacetylene. Experimentally, the
compound crystallizes with 1 mol of cyclohexane. A molecular model of
207 with all methyls turned inward is complementary to chair cyclohexane. A model with all methyls turned outward is complementary to a
model of ferrocene. In the model of the caviplex, each C–H bond of the
guest contacts an oxygen of the host. Although **206** and **207** contain
enforced cavities of molecular dimensions (cavitands), their interiors are
adaptable to a variety of potential guests. Thus, the host can provide
the guest with contact with either methyl groups or with the oxygen's
unshared electron pairs, depending on the electronic and geometric
demands of the guests. These features remain to be exploited.[180] The
next steps in research on these cavitands is to append to them water-
solubilizing and catalytic groups. The former will provide them with
hydrophobic driving forces to complex nonpolar guests and the latter to
catalyze reactions of such guests.

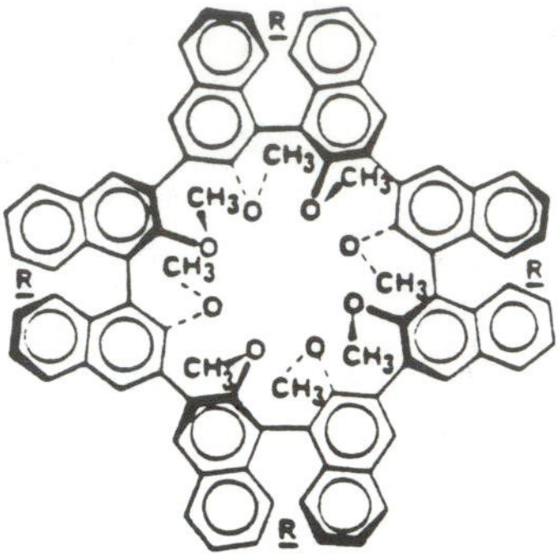

206

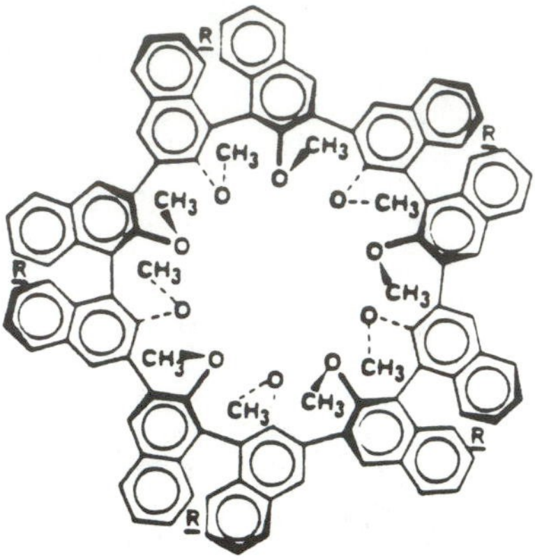

207

Carcerands—Synthetic Molecular Cells

Absent among the millions of organic compounds hitherto reported are closed-surface hosts with enforced interiors large enough to "imprison," behind covalent "bars," guests the size of ordinary solvent molecules. After much thought and molecular model examination, compound **210** became the target for synthesis of the first molecular cell. The term carcerand was applied to this class of compound ("Cram's test"). The synthesis, by S. Karbach, Y. H. Kim, et al.[181] involved treating Cs_2CO_3 with a solution in $(CH_3)_2NCHO-(CH_2)_4O$ of equal molar amounts of cavi-

At the dinner for the first holder of the Saul Winstein Chair at UCLA. Left to right, Carolee Winstein, Donald J. Cram, and Edith Roberts on June 2, 1980. Photo lent by Carolee Winstein.

tands **208** and **209** under an atmosphere of argon. The first question to be answered was "What guest compounds would be trapped inside during the shell closure?" This question is akin to asking whether two soup bowls closed rim-to-rim under the surface of a kettle of stew would net any stew. The answer was that **210** "contained" essentially every component of the medium present during ring closure.

The products (**210** and guests) were very insoluble in all media, and **210** was purified by extracting it with the most powerful solvents of each type. The remaining material was subjected to elemental analysis for C, H, S, O, N, Cl, and Cs. Nitrogen analysis and an IR spectrum of the substance revealed that $(CH_3)_2NCHO$ had been entrapped. The

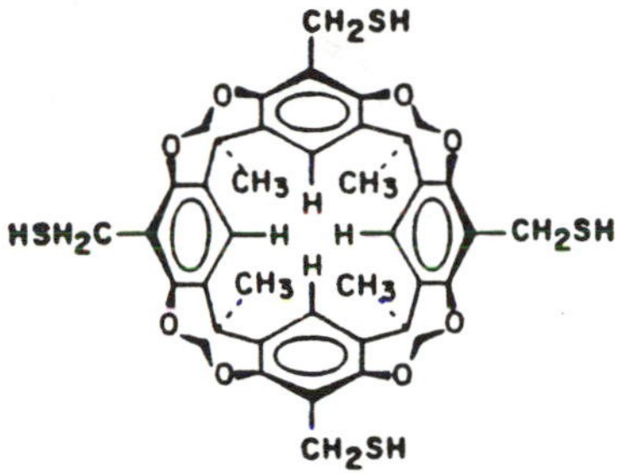

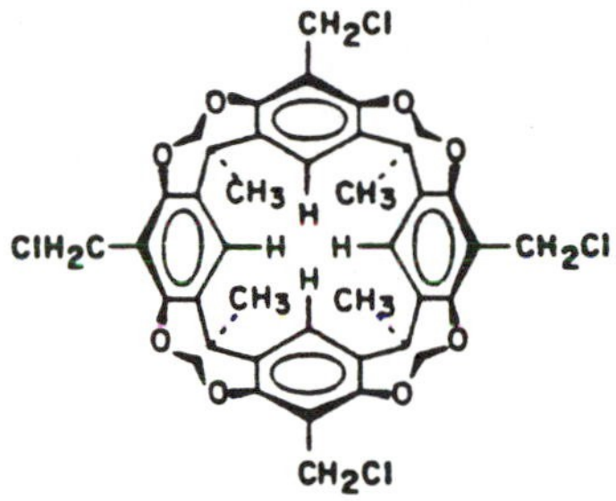

208   **209**

$$\frac{(CH_3)_2NCHO-(CH_2)_4O}{Cs_2CO_3, \ Ar} \longrightarrow$$

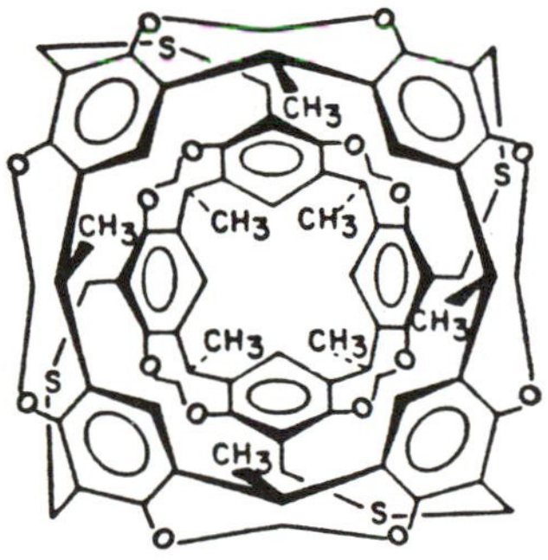

210

presence of equivalent amounts of Cs and Cl demonstrated that one or the other ion or both had to be encapsulated in the host.

A fast atom bombardment mass spectrum (FABMS) of **210·G** (G stands for guest) showed the presence of the following host–guest combinations. The species trapped in the interior of **210** are enclosed in parentheses.

210 · no guest   **210** · (Cs^+) · Cl^-

210 · $((CH_3)_2NCHO))$ **210** · $(Cs^+ + H_2O)$ · Cl^-

210 · $((CH_2)_4O + H_2O)$ **210** · $((CH_3)_2NCHO + Cs^+)$ · Cl^-

210 · $(Cs^+ + Ar)$ · Cl^- **210** · $(Cs^+ + H_2O + Cs^+)$ · Cl_2^{2-}

210 · $(Cs^+ + Cl^-)$ **210** · $(Cs^+ + Cs^+ + Cl^-)$ · Cl^-

No peaks were found at molecular masses above that of the last carcaplex listed. None were observed that could not be interpreted in terms of appropriate host–guest combinations. When highly dried **210** · G was boiled with D_2O, the **210** · $(Cs^+ + H_2O)$ peak was substantially replaced by a **210** · $(Cs^+ + D_2O)$ peak. Models suggest that **210** has two small portals lined with methyl groups through which molecules as small as H_2O can pass.

Molecular models of **210** show that its interior surface is complementary to the outer surface of *anti*-$ClCF_2CF_2Cl$. Shell closure of **208** and **209** in the presence of this chlorofluorocarbon (Freon) resulted in entrapment of a small amount of Freon in the interior of **210**.

The FABMS coupled with the elemental analyses indicated that about 5% of the mixture was noncomplexed **210**, about 60% was encapsulated Cs^+, about 45% was encapsulated $(CH_3)_2NCHO$, 15% was encapsulated $(CH_2)_4O$, but only 1–2% was encapsulated Cl^-. Thus, Cs^+ was mainly inside and Cl^- was mainly outside the carcaplex. Models show that if the final covalent bond leading to **210** · G involved an intramolecular S_N2 linear transition state, as in **211**, any Cs^+ ion-paired to the S^- is trapped inside the cavity and the Cl^- must be external to the cavity.[181] I anticipate that unusual physical and chemical properties will provide unusual uses for carcaplexes, particularly when their design renders them soluble and separable.

$$Cs^+S^- \overbrace{\qquad} CH_2 \longrightarrow Cl$$

$\underset{\sim\,\sim\,\sim}{211}$

Organic Textbook Writing and Teaching

While attending Harvard (1945–1947), George Hammond and I planned to coauthor a basic textbook of organic chemistry whose organic reactions were classified according to their mechanisms, rather than according to the structures of their starting materials or products. To our mutual discomfort, we struggled hard over the minor points and style of the book but little over the main themes. In 1959, after 2 years of battle and compromise, the book was published.[182] Like most "first books", this one was written more for the authors than for their readers. We wished to demonstrate to ourselves that we could write a better textbook with a better organization than anyone else. This endeavor provided me with a broad view of organic chemistry that was useful in my research, teaching, and consulting. It also gave me a profound respect for George Hammond. After completing its first draft, I immediately wrote 17 research papers to prove that textbook writing had not been substituted for carrying on research.

This book, *Organic Chemistry*, has enjoyed four American and four international student editions. One or another of the editions has been translated into 13 different languages. We were joined as authors in the third edition by James B. Hendrickson, and in the fourth edition by Stanley H. Pine.

Since coming to UCLA in 1947, I estimate that I have taught second-year organic chemistry to about 10,000 undergraduate students, most of whom were preparing themselves for careers in the life and physical sciences. My wife, Jane M. Cram, taught a similar course at Mount Holyoke College on and off for 15 years. During the 1950s and 1960s we watched biochemistry, physiology, and genetics being reduced to organic and physical organic concepts. Following the dictum, "If you want to learn something, teach it", we decided to write an elementary organic textbook for life science students that dealt only with those aspects of organic chemistry useful to life science students. We drew as many parallels as possible between laboratory and biological chemistry in our book *The Essence of Organic Chemistry*.[183]

The main benefit of coauthoring a book is the criticism that flows back and forth between authors. The book, more than the authors, enjoyed this benefit. Interestingly, the Japanese translation of the book was more widely adopted for classes in Japan than the American edition was for classes in the United States. These books have not been without their influence on the thinking of scientists during their formative periods. The blending of physical, inorganic, and biochemical with organic chemical concepts grows ever more important. From 1970 on, my research program was devoted to biomimetic problems.

While on a consulting trip to Upjohn, I injured my back playing tennis to the extent that I knew it would be impossible for me to stand at a board and lecture for an hour. Because I was scheduled to teach an elementary organic chemistry course to about 250 second-year students on the day after I returned to Los Angeles, I prevailed upon my wife, Jane, to assist me. Because she had taught organic chemistry at Mt. Holyoke for 15 years, she was well-qualified to work as my assistant. I delivered all of my lectures from a chair in the lecture room while Jane wrote the formulas on the board. After about 7 weeks of this "team teaching", my back mended to the point that I could teach the course unaided. The students apparently liked the "Ma" and "Pa" part of the course much better than my solo part. One student remarked, "Your wife's formulas were much more slowly and carefully written than yours", and another said, "We enjoyed the interchanges between you and your wife, particularly those you didn't think we heard."

Teaching and research are natural and happy companions. In chemistry, classroom teaching deals with the research results of only about 150 years, whereas research is finding out what will be taught in the future. Teaching stimulates the process of discovering research problems and of turning research results into an explicit, communicable form. Research generates enthusiasm and an appetite to tell others about results. Teaching graduate students and postdoctoral scholars how to think about and do research is tutorial and great fun. I will have had about 118 graduate students carry on their Ph.D. thesis research with me when those currently in my group graduate. About 87 postdoctoral scholars from 21 different countries have worked with me. My research group size has averaged 16 to 18 co-workers over the years.

My upbringing was much influenced by the idealized image of my father painted by my mother. For example, she told me he gave up smoking when I was born. I have never smoked. She said that he always kept himself physically fit and loved outside participatory sports. My attitude has been that physical games made way for mental games involving research. The kinesthetic sports of skiing, tennis, swimming, sailing, and surfboarding have cleansed my mind on occasions beyond mention of the cobwebs of confusion, doubt, and pessimism that are the unwelcome companions of the gambling research game. One of my favorite thoughts has been, "I have only one mind and body and must live in them for my whole life." The keeping of my mind clear and flexible has depended heavily on keeping my muscles resilient and used. The violence of fighting my way out through the "herds of white buffaloes"—slang for large broken surfing waves—has had a civilizing and moderating influence on my professional life.

In retrospect, I judge that my father's death just before I was 4 years old forced me to construct a model for my own character that was composed of pieces taken from many different individuals; some being people I studied and others I lifted from books. Bad examples were

Jane Cram and Donald on a ski lift at Snow Mass, Colorado, about 1978.

much more abundant, and taken in sum, were at least as important as the good examples mentioned in this biography. My early reading of essentially all of the works of Dickens, Scott, Shaw, Kipling, and some of Shakespeare, Dostoevski, Spenger, and Tolstoy left me with a generally romantic and optimistic attitude toward life and people. My character and sense of values have formed slowly, and even at 68 years of

Donald J. Cram and M. Tkach sailing on La Valera *in 1967 near Long Beach, California.*

Donald J. Cram riding a big one at San Onofre, 1988.

age, are still developing. For example, Robert Ardrey's comparisons of human and animal behavior (1961–1970) allowed me to integrate many experiences. As a youth, I could not understand why society and its institutions were so disorderly. Now, as someone of late middle age, I am amazed that, given the diversity of society, it is able to function at all.

While a senior at Rollins College in 1940, I applied to 17 universities for admission to graduate school as a teaching assistant. Three offers materialized, the best of which was the University of Nebraska ($50 per month) where I obtained an M.S. degree in 11 months. I greatly enjoyed my thesis work (1942) under the supervision of Norman H. Cromwell. I have since lectured at most of the 14 schools that turned me down.

Many people have influenced me: Guy Waddington at Rollins College, Max Tishler at Merck, Paul Bartlett and Robert Woodward at Harvard, and William Young and Saul Winstein at UCLA. Each, in special ways, provided me with examples of excellence. I have much admired Vlado Prelog, my international colleague–competitor for some 35 years, and am indebted to John D. Roberts and George S. Hammond for helping me learn how to think about research problems over a 40-year period.

My early hopes of avoiding repetition and working in a profession that challenged my creative and organizational capacities have been fulfilled beyond my expectations. The opportunities for being creative in organic chemical research are open-ended. Unlimited possibilities exist to design new compounds, reactions, synthetic sequences, and test systems and to discover new principles, correlations, and uses. I know of no other profession as rich in diverse phenomena and scientific enterprise as that of the organic research chemist. No other kind of scientist engages in such a variety of types of scientific reasoning and in putting predictions to test with such speed. The amalgams of certainty and uncertainty, of the approximate with the exact, of speculation and observable result have been my daily companions for over 4 decades.

Among its many rewarding aspects, academic chemical research has a vast literature, each article of which represents a piece of some scientist's biography. This "keeping track" is enormously satisfying to me. I like to know what has and what has not been done. Any chemists reading this book can see, in some detail, how I have spent most of my mature life. They can become familiar with the quality of my mind and imagination. They can make judgments about my research abilities. They can tell how well I have documented my claims of experimental results. Any scientist can redo my experiments to see if they still work—and this has happened! I know of no other field in which contributions to world culture are so clearly on exhibit, so cumulative, and so subject to verification.

The California Museum of Science and Industry presents a $5000 award each year to the "California Scientist of the Year", who is selected by a committee of outstanding scientists. This award, along with an award to the outstanding industrialist of the year, is given at a black tie dinner attended by about 500 of the Los Angeles "elite." I was fortunate enough to be selected "California Scientist of the Year" in 1974. I had attended several such dinners in previous years, in which various scientists gave accounts of their successes. In an attempt to avoid the boredom of an autobiography, I wrote and sang the following ballad at the banquet as my acceptance speech. This enlivened the occasion, and was warmly approved by the audience (with the exception of certain scientists who thought that science should be above and beyond fun and humor). I have never thought that *science* was funny, but have always regarded *scientists* as one grade more humorous than undertakers.

Autobiographical Parody on the Career
of the 1974 California Scientist of the Year

(with apologies to the captain of the *H.M.S. Pinafore*
and to those of you who have taste in music)

When I was a boy I learned the rule,
To avoid hard work I must go to school,
I learned to add and subtract with ease,
To keep my grades high in the As and Bs.
I learned these lessons so thoroughly,
That passing courses was a cinch for me,
I learned these lessons so carefully—
I passed these courses just masterfully.

I played ice hockey and football too,
Where I found that bigness did a lot for you.
I learned the lesson that to play first string—
You worked and practiced at everything.
I worked and practiced so faithfully,
My hopes were to get myself a college degree.
I worked and practiced so faithfully,
I hoped to get myself a college degree.

In high school in my senior year
Came the turning point of my career.
I took a course in chemistry,
And found that I handled it proficiently.
My interest in the science did so persist,
I decided on becoming a scientist—
My interest in the subject did so persist,
I made sure that I would become a scientist.

In my first college lab course, I broke so much glass,
That I found myself at the bottom of the class.
But I learned how to filter and crystallize—
And to separate, identify, and analyze.
I learned how to crystallize so skillfully,
I hoped to attend a university—
I learned how to crystallize so expertly,
I matriculated at a university.

In graduate school I discovered soon—
That chemical research provided lots of room.
To do experiments that had never been done—
That to get results was lots of fun.
I synthesized compounds so prolifically,
That Harvard granted me a Ph.D.
I synthesized compounds so rapidly,
That Harvard granted me a Ph.D.

To UCLA I took a trip,
To fill an appointment of a fellowship.
To conduct research and find what's new,
To prospect, experiment, and see what's true.
I wrote and I published with so much flair,
My colleagues voted a professor's chair.
I wrote and I published with so much flair,
The regents approved that professor's chair.

From fifteen countries my students came,
To gamble and win in the research game.
With synthetic complexes they demonstrated,
That nature's methods could be imitated.
Their hard work brought to my career,
This honor of Scientist of the Year—
Their labors brought to my career—
The California Scientist of the Year.

In 1987, J.-M. Lehn, Charles Pedersen, and I shared the Nobel prize in chemistry. The most immediate pleasure of receiving the prize was seeing the joy and excitement of my co-workers, friends, students, and colleagues. Congratulatory mail, cables, and telegrams poured in, and my wife and I were lavished with much attention by the news media. Many invitations to lecture and to address various groups were received and continue to come in. I realized it was debt-paying time to my alma maters, the University, my co-workers, my colleagues, and my friends. In a period of 3 weeks, with little preparation, I addressed the Board of Regents of the University of California, the UCLA Senate, three fund-raising foundations at UCLA, and the Los Angeles County Board of Supervisors. Then, my wife, Jane, and I were off to Sweden for the Nobel ceremonies that were nonstop for 9 days.

To me, the Nobel prize is a celebration of the spirit of inquiry. Everything we know and use was at one time or another the object of inquiry and investigation. My getting this award at the age of 68, although temporarily diverting, is likely to extend my research career, for which I am very grateful. Besides, who doesn't like to be honored?

The interesting question arises as to whether there is enough unity in my research, conducted over a 45-year period, to allow it to be characterized. In my opinion, organic chemists are part artists and part scientists, and thus apply both lobes of their brains to their work. To the extent that they are artists, they develop a research style expressed in their choice of research problems, how they address these problems, the degree of craftsmanship they bring to their research results, the extent to which they document their results, the readership they address in their papers, and their style of writing papers.

I have always felt that I understood a phenomenon only to the extent that I could visualize it. Much of the charm organic chemical research has for me derives from structural formulas. When reading chemical journals, I look for formulas first. Very few of the nearly 370 papers I have published are without structural formulas.

The chronology of the different fields of research I have worked in is as follows: isolation and determination of the structures of natural products; application of stereochemical probes to organic reaction mechanisms (e.g., phenonium ions, elimination reactions, Cram's rule, carbanion reactions, ion-pair reorganization reactions, and substitution reactions at sulfur); cyclophane chemistry; and host—guest complexation chemistry. Chemical architecture and manipulation of the symmetry properties of molecules have been dominant themes in the research in each of these fields. Other themes also evident are the documentation of research results by publication of full-length papers, a

John D. Roberts and Donald J. Cram at the reception following announcement of the Nobel prize, October 1987.

cadence and rhythm of writing, the use of favorite expressions and words. I have just reread my papers on the phenonium ion and instantly recognized my style of expression in sentences and paragraphs, even though these papers were written about 38 years ago.

Acknowledgments

I feel privileged to live in a society that has faith in research and that supports it generously. I thank the granting agencies and those that administer them. In particular, the Division of Basic Energy Sciences of the Department of Energy supported the research on the metal ion binding by the spherands, hemispherands, and cryptaspherands; the National Science Foundation supported the research on chiral recognition, cavitands, and carcerands. The National Institutes of Health supported the work on catalysis through complexation. I warmly thank all former and present members of my research group and the many others whose results and discussions have stimulated and instructed us over the years. The artwork displayed in this volume was done by June Hendrix, to whom I am much indebted.

References

1. Cram, D. J. *J. Am. Chem. Soc.* **1948,** *70,* 4240–4243.

2. Cram, D. J.; Tishler, M. *J. Am. Chem. Soc.* **1948,** *70,* 4238–4239.

3. Cram, D. J. *J. Am. Chem. Soc.* **1948,** *70,* 4244–4247.

4. Cram, D. J. *J. Am. Chem. Soc.* **1949,** *71,* 3863–3870.

5. Cram, D. J. *J. Am. Chem. Soc.* **1949,** *71,* 3875–3883.

6. Cram, D. J. *J. Am. Chem. Soc.* **1952,** *74,* 2129–2137.

7. Winstein, S.; Trifan, D. *J. Am. Chem. Soc.* **1952,** *74,* 1147–1154.

8. Winstein, S.; Trifan, D. *J. Am. Chem. Soc.* **1952,** *74,* 1154–1160.

9. Lancelot, C.; Cram, D. J.; Schleyer, P. R. In *Carbonium Ions;* Olah, G. A.; Schleyer, P. R., Eds.; Interscience: New York, 1972; Chapter 27, Vol. 3, pp 1347–1483.

10. Cram, D. J.; Elhafez, F. A. A. *J. Am. Chem. Soc.* **1952,** *74,* 5828–5835.

11. Cram, D. J. *J. Am. Chem. Soc.* **1949,** *71,* 3883–3889.

12. Huckel, W.; Tapp, W. *Legutke G. Ann.* **1940,** *543,* 191–230.

13. Cram, D. J.; McCarty, J. E. *J. Am. Chem. Soc.* **1954,** *76,* 5740–5745.

14. Cram, D. J.; Sahyun, M. R. V. *J. Am. Chem. Soc.* **1963,** *85,* 1263–1268.

15. Kingsbury, C. A.; Cram, D. J. *J. Am. Chem. Soc.* **1960,** *82,* 1810–1819.

16. Cram, D. J.; Greene, F. D.; DePuy, C. H. *J. Am. Chem. Soc.* **1956**, *78*, 790–796.

17. Dewar, M. J. S. *Nature (London)* **1945**, *156*, 784.

18. Cram, D. J.; Steinberg, H. *J. Am. Chem. Soc.* **1951**, *73*, 5691–5704.

19. Brown, C. J.; Farthing, A. C. *Nature (London)* **1949**, *164*, 915–916.

20. Hope, H.; Bernstein, J.; Trueblood, K. N. *Acta Cryst.* **1972**, *B28*, 1733–1743.

21. Reichstein, T.; Oppenauer, R. *Helv. Chim. Acta* **1933**, *16*, 1373–1380.

22. Baker, W.; McOmie, J. F. W.; Norman, J. M. *Chem. Ind. (London)* **1950**, 77.

23. Cram, D. J.; Abell, J. *J. Am. Chem. Soc.* **1955**, *77*, 1179–1186.

24. Winberg, H. E.; Fawcett, F. S.; Mochel, W. E.; Theobald, C. W. *J. Am. Chem. Soc.* **1960**, *82*, 1428–1435.

25. Cram, D. J.; Allinger, N. L.; Steinberg, H. *J. Am. Chem. Soc.* **1954**, *76*, 6132–6141.

26. Cram, D. J.; Helgeson, R. C. *J. Am. Chem. Soc.* **1966**, *88*, 3515–3521.

27. Cram, D. J.; Helgeson, R. C.; Lock, D.; Singer, L. A. *J. Am. Chem. Soc.* **1966**, *88*, 1324–1325.

28. Delton, M. H.; Cram, D. J. *J. Am. Chem. Soc.* **1972**, *94*, 1669–1675.

29. Gilman, R. E.; Delton, M. H.; Cram, D. J. *J. Am. Chem. Soc.* **1972**, *94*, 2478–2482.

30. Cram, D. J.; Montgomery, C. S.; Knox, G. R. *J. Am. Chem. Soc.* **1966**, *88*, 515–525.

31. Allinger, N. L.; Freiberg, L. H.; Hermann, R. B.; Miller, M. H. *J. Am. Chem. Soc.* **1963**, *85*, 1171–1176.

32. Trueblood, K. N.; Crisp, M. J., private communication; drawn in Cram, D. J.; Cram, J. M. *Acc. Chem. Res.* **1971**, *4*, 204–213.

33. Cram, D. J.; Bauer, R. H.; Allinger, N. L.; Reeves, R. A.; Wechter, W. J.; Heilbronner, E. *J. Am. Chem. Soc.* **1959**, *81*, 5977–5983.

34. Kammula, S. L.; Iroff, L. D.; Jones, M., Jr.; van Straten, J. W.; deWolf, W. H.; Bickelhaupt, F. *J. Am. Chem. Soc.* **1977**, *99*, 5815.

35. Cram, D. J.; Bauer, R. H. *J. Am. Chem. Soc.* **1959**, *81*, 5971–5977.

36. Singer, L. A.; Cram, D. J. *J. Am. Chem. Soc.* **1963**, *85*, 1080–1084.

37. Reich, H. J.; Cram, D. J. *J. Am. Chem. Soc.* **1969,** *91,* 3505–3516.

38. Truesdale, E. A.; Cram, D. J. *J. Org. Chem.* **1980,** *45,* 3974–3981.

39. Shieh, C.; McNally, D. C.; Boyd, R. H. *Tetrahedron* 1969, *25,* 3653–3665.

40. Reich, H. J.; Cram, D. J. *J. Am. Chem. Soc.* **1969,** *91,* 3517–3526.

41. Cram, D. J.; Wechter, W. J.; Kierstead, R. W. *J. Am. Chem. Soc.* 1958, *80,* 3126–3132.

42. Cram, D. J.; Hefelfinger, D. T. *J. Am. Chem. Soc.* **1971,** *93,* 4754–4772.

43. Dewhirst, K. C.; Cram, D. J. *J. Am. Chem. Soc.* **1958,** *80,* 3115–3125.

44. Dewhirst, K. C.; Cram, D. J. *J. Am. Chem. Soc.* **1959,** *81,* 5963–5971.

45. Otsubo, T.; Gray, R.; Boekelheide, V. *J. Am. Chem. Soc.* **1978,** *100,* 2449–2456.

46. *Cyclophanes;* Keehn, P. M.; Rosenfeld, S. M., Eds.; Academic: New York, 1983; Vol. 1, pp 1–357; Vol. 2, pp 359–725.

47. Cram, D. J. *Fundamentals of Carbanion Chemistry;* Academic: New York, 1965; pp 1–289.

48. Wilson, C. L. *J. Chem. Soc.* **1936,** 1550–1553.

49. Hsu, S. K.; Ingold, C. K.; Wilson, C. L. *J. Chem. Soc.* **1938,** 78–81.

50. Cram, D. J.; Allinger, J.; Langemann, A. *Chem. Ind. (London)* 1955, 919–920.

51. Cram, D. J.; Mateos, J. L.; Hauck, F.; Langemann, A.; Kopecky, K. R.; Nielsen, W. D.; Allinger, J. *J. Am. Chem. Soc.* **1959,** *81,* 5774–5784.

52. Cram, D. J.; Bradshaw, J. S. *J. Am. Chem. Soc.* **1963,** *85,* 1108–1118.

53. Cram, D. J; Kingsbury, C. A.; Langemann, A. *J. Am. Chem. Soc.* **1959,** *81,* 5785–5790.

54. Cram, D. J.; Kingsbury, C. A.; Rickborn, B. *J. Am. Chem. Soc.* 1959, *81,* 5835.

55. Cram, D. J.; Rickborn, B.; Kingsbury, C. A.; Haberfield, P. *J. Am. Chem. Soc.* **1961,** *83,* 3678–3687.

56. Cram, D. J.; Kingsbury, C. A.; Rickborn, B. *J. Am. Chem. Soc.* 1961, *83,* 3688–3696.

57. Cram, D. J.; Gosser, L. *J. Am. Chem. Soc.* **1964,** *86,* 5445–5457.

58. Cram, D. J.; Gosser, L. *J. Am. Chem. Soc.* **1964,** *86,* 2950–2952.

59. Cram, D. J.; Ford, W. T.; Gosser, L. *J. Am. Chem. Soc.* **1968,** *90,* 2598–2606.

60. Ford, W. T.; Graham, E. W.; Cram, D. J. *J. Am. Chem. Soc.* **1967,** *89,* 4661–4669.

61. Cram, D. J.; Gosser, L. *J. Am. Chem. Soc.* **1964,** *86,* 5457–5465.

62. Cram, D. J.; Nielsen, W. D.; Rickborn, B. *J. Am. Chem. Soc.* **1960,** *82,* 6415.

63. Cram, D. J.; Scott, D. A.; Nielsen, W. D. *J. Am. Chem. Soc.* **1961,** *83,* 3696–3707.

64. Cram, D. J.; Uyeda, R. T. *J. Am. Chem. Soc.* **1962,** *84,* 4358.

65. Bergson, G.; Weidler, A. M. *Acta Chem. Scand.* **1963,** *17,* 862–864; 1798–1799; 2691–2700; 2724–2734.

66. Bank, S.; Rowe, C. A., Jr.; Schriesheim, A. *J. Am. Chem. Soc.* **1963,** *85,* 2115–2118.

67. Bates, R. B.; Carnighan, R. H.; Staples, C. E. *J. Am. Chem. Soc.* **1963,** *85,* 3032–3033.

68. Doering, W. von E.; Gaspar, P. P. *J. Am. Chem. Soc.* **1963,** *85,* 3043.

69. Kawahara, F. S.; Talalay, P. *J. Biol. Chem.* **1960,** *235,* PC 1–2.

70. Agranoff, B. W.: Eggerer, H.; Henning U.; Lynen, F. *J. Biol. Chem.* **1960,** *235,* 326–332.

71. Rilling, H. C.; Coon, M. J. *J. Biol. Chem.* **1960,** *235,* 3087–3092.

72. Cram, D. J.; Cram, J. M. *Intra-Sci. Chem. Rep.* **1973,** *7,* 1–17.

73. Cram, D. J.; Willey, F.; Fischer, H. P.; Scott, D. A. *J. Am. Chem. Soc.* **1964,** *86,* 5370–5371.

74. Cram, D. J.; Willey, F.; Fischer, H. P.; Relles, H. M.; Scott, D. A. *J. Am. Chem. Soc.* **1966,** *88,* 2759–2766.

75. Almy, J.; Uyeda, R. T.; Cram, D. J. *J. Am. Chem. Soc.* **1967,** *89,* 6768–6770.

76. Almy, J.; Cram, D. J. *J. Am. Chem. Soc.* **1969,** *91,* 4459–4468.

77. Almy, J.; Garwood, D. C.; Cram, D. J. *J. Am. Chem. Soc.* **1970,** *92,* 4321–4330.

78. Almy, J.; Hoffman, D. H.; Chu, K. C.; Cram, D. J. *J. Am. Chem. Soc.* **1973**, *95*, 1185–1190.

79. Ingold, C. K. *Structure and Mechanism in Organic Chemistry*; Cornell University Press: Ithaca, NY, 1953, p 572.

80. Cram, D. J.; Guthrie, R. D. *J. Am. Chem. Soc.* **1966**, *88*, 5760–5765.

81. Guthrie, R. D.; Jaeger, D. A.; Meister, W.; Cram, D. J. *J. Am. Chem. Soc.* **1971**, *93*, 5137–5153.

82. Jaeger, D. A.; Cram, D. J. *J. Am. Chem. Soc.* **1971**, *93*, 5153–5161.

83. Jaeger, D. A.; Broadhurst, M. D.; Cram, D. J. *J. Am. Chem. Soc.* **1979**, *101* 717–732.

84. Corey, E. J.; Kaiser, E. T. *J. Am. Chem. Soc.* **1961**, *83*, 490–491.

85. Goering, H. L.; Towns, D. L.; Dittmer, B. *J. Org. Chem.* **1962**, *27*, 736–739.

86. Cram, D. J.; Wingrove, A. S. *J. Am. Chem. Soc.* **1962**, *84*, 1496.

87. Cram, D. J.; Wingrove, A. S. *J. Am. Chem. Soc.* **1963**, *85*, 1100–1107.

88. Cram, D. J.; Partos, R. D. *J. Am. Chem. Soc.* **1963**, *85*, 1093–1096.

89. Cram, D. J.; Pine, S. H. *J. Am. Chem. Soc.* **1963**, *85*, 1096–1100.

90. Cram, D. J.; Trepka, R. D.; St. Janiak, P. *J. Am. Chem. Soc.* **1966**, *88*, 2749–2759.

91. Corey, E. J.; Konig, H.; Lowry, T. H. *Tetrahedron Lett.* **1962**, 515–520.

92. Corey, E. J.; Lowry, T. H. *Tetrahedron Lett.* **1965**, 793–801.

93. Corey, E. J.; Lowry, T. H. *Tetrahedron Lett.* **1965**, 803–809.

94. Bordwell, F. G.; Phillips, D. D.; Williams, J. M., Jr. *J. Am. Chem. Soc.* **1968**, *90*, 426–428.

95. Bordwell, F. G.; Doomes, E.; Corfield, P. W. R. *J. Am. Chem. Soc.* **1970**, *92*, 2581–2583.

96. Wolfe, S.; Rauk, A.; Csizmadia, I. G. *J. Am. Chem. Soc.* **1969**, *91*, 1567–1569.

97. Cram, D. J.; Rickborn, B.; Knox, G. R. *J. Am. Chem. Soc.* **1960**, *82*, 6412.

98. Stewart, R.; O'Donnell, J. P. *J. Am. Chem. Soc.* **1962**, *84*, 493–494.

99. Stewart, R.; O'Donnell, J. P.; Cram, D. J.; Rickborn, B. *Tetrahedron* **1962,** *18,* 917–922.

100. Conant, J. B.; Wheland, G. W. *J. Am. Chem. Soc.* **1932,** *54,* 1212–1221.

101. McEwen, W. K. *J. Am. Chem. Soc.* **1936,** *58,* 1124–1129.

102. Pedersen, C. J. *J. Am. Chem. Soc.* **1967,** *89,* 2495–2496.

103. Pedersen, C. J. *J. Am. Chem. Soc.* **1967,** *89,* 7017–7036.

104. Dietrich, B.; Lehn, J.–M.; Sauvage, J.–P. *Tetrahedron Lett.* **1969,** 2885–2888.

105. Dietrich, B.; Lehn, J.–M.; Sauvage, J.–P. *Tetrahedron Lett.* **1969,** 2889–2892.

106. Kyba, E. P.; Siegel, M. G.; Sousa, L. R.; Sogah, G. D. Y.; Cram, D. J. *J. Am. Chem. Soc.* **1973,** *95,* 2691–2692.

107. Kyba, E. P.; Koga, K.; Sousa, L. R.; Siegel, M. G.; Cram, D. J. *J. Am. Chem. Soc.* **1973,** *95,* 2692–2693.

108. Helgeson, R. C.; Koga, K.; Timko, J. M.; Cram, D. J. *J. Am. Chem. Soc.* **1973,** *95,* 3021–3023.

109. Helgeson, R. C.; Timko, J. M.; Cram, D. J. *J. Am. Chem. Soc.* **1973,** *95,* 3023–3025.

110. Gokel, G. W.; Cram, D. J. *J. Chem. Soc. Chem. Commun.* **1973,** *521,* 481–482.

111. Cram, D. J.; Cram, J. M. *Science (Washington, DC)* **1974,** *183,* 803–809.

112. Kyba, E. P.; Helgeson, R. C.; Madan, K.; Gokel, G. W.; Tarnowski, T. L.; Moore, S. S.; Cram, D. J. *J. Am. Chem. Soc.* **1977,** *99,* 2564–2571.

113. Koltun, W. L. *Biopolymers* **1965,** 3, 665–679.

114. The crystal structures and references to them up to 1980 are gathered in Cram, D. J.; Trueblood, K. N. *Top. Curr. Chem.* **1981,** *98,* 43–106.

115. Newcomb, M.; Moore, S. S.; Cram, D. J. *J. Am. Chem. Soc.* **1977,** *99,* 6405–6410.

116. Bell, T. W.; Cheng, P. G.; Newcomb, M.; Cram, D. J. *J. Am. Chem. Soc.* **1982,** *104,* 5185–5188.

117. Knobler, C. B.; Trueblood, K. N.; Weiss, R. M., private communication.

118. Timko, J. M.; Moore, S. S.; Walba, D. M.; Hiberty, P. C.; Cram, D. J. *J. Am. Chem. Soc.* **1977**, *99*, 4207–4219.

119. Goldberg, I. *J. Am. Chem. Soc.* **1980**, *102*, 4106–4113.

120. Koenig, K. E.; Lein, G. M; Stuckler, P.; Kaneda, T.; Cram, D. J. *J. Am. Chem. Soc.* **1979**, *101*, 3553–3566.

121. Helgeson, R. C.; Tarnowski, T. L.; Cram, D. J. *J. Org. Chem.* **1979**, *44*, 2538–2550.

122. Knobler, C. B.; Maverick, E. F.; Trueblood, K. N.; Helgeson, R. C.; Cram, D. J. *Acta Cryst.* **1986**, *C42*, 156–158.

123. Doxsee, K. M.; Cram, D. J. *J. Org. Chem.* **1986**, *51*, 5068–5071.

124. Haymore, B. L.; Huffman, J. C., 4th Symposium on Macrocyclic Compounds, Provo, UT, August 1980.

125. deBoer, J. A. A.; Uiterwijk, J. W. H. M.; Geevers, J.; Harkema, S.; Reinhoudt, D. N. *J. Org. Chem.* **1983**, *48*, 4821–4830.

126. Peacock, S. C.; Domeier, L. A.; Gaeta, F. C. A.; Helgeson, R. C.; Timko, J. M.; Cram, D. J. *J. Am. Chem. Soc.* **1978**, *100*, 8190–8202.

127. Peacock, S. C.; Walba, D. M.; Gaeta, F. C. A.; Helgeson, R. C.; Cram, D. J. *J. Am. Chem. Soc.* **1980**, *102*, 2043–2052.

128. Newcomb, M.; Toner, J. L.; Helgeson, R. C.; Cram, D. J. *J. Am. Chem. Soc.* **1979**, *101*, 4941–4947.

129. Sogah, G. D. Y.; Cram, D. J. *J. Am. Chem. Soc.* **1979**, *101*, 3035–3042.

130. Lingenfelter, D. S.; Helgeson, R. C.; Cram, D. J. *J. Org. Chem.* **1981**, *46*, 393–406.

131. Langstrom, B.; Bergson, G. *Acta Chem. Scand.* **1973**, *27*, 3118–3132.

132. Wynberg, H.; Hermann, K. *J. Org. Chem.* **1979**, *44*, 2238–2244.

133. Cram, D. J.; Sogah, G. D. Y. *J. Chem. Soc. Chem. Commun.* **1981**, 625–628.

134. Mazaleyrat, J.–P.; Cram, D. J. *J. Am. Chem. Soc.* **1981**, *103*, 4585–4586.

135. Seebach, D.; Crass, G.; Wilka, E. M.; Hilvert, D.; Brunner, E. *Helv. Chim. Acta* **1979**, *62*, 2695–2698.

136. Mukaiyama, T.; Soai, K.; Sato, T.; Shimizu, H.; Suzucki, K. *J. Am. Chem. Soc.* **1979**, *101*, 1455–1460.

137. Okamoto, Y.; Suzucki, K.; Ohta, K.; Hatada, K.; Yuki, H. *J. Am. Chem. Soc.* **1979**, *101*, 4763–4765.

138. Drenth, W.; Nolte, R. J. M. *Acc. Chem. Res.* **1979**, *12*, 30–35.

139. Cram, D. J.; Sogah, G. D. Y. *J. Am. Chem. Soc.* **1985**, *107*, 8301–8302.

140. Dunitz, J. D.; Dobler, M.; Seiler, P.; Phizackerly, R. P. *Acta Crystallogr. Sect. B* **1974**, *30*, 2733, and following papers to 2750.

141. Weiss, R.; Metz, B.; Moras, D. *Proc. Int. Conf. Coord. Chem. 13th.* **1970**, *2*, 85–86.

142. Metz, B.; Moras, D.; Weiss, R. *Acta Crystallogr. Sect. B* **1973**, *29*, 1377–1381.

143. Cram, D. J.; Kaneda, T.; Helgeson, R. C.; Lein, G. M. *J. Am. Chem. Soc.* **1979**, *101*, 6752–6754.

144. Trueblood, K. N.; Knobler, C. B.; Maverick, E.; Helgeson, R. C.; Brown, S. B.; Cram, D. J. *J. Am. Chem. Soc.* **1981**, *103*, 5594–5596.

145. Helgeson, R. C.; Weisman, G. R.; Toner, J. L.; Tarnowski, T. L.; Chao, Y.; Mayer, J. M.; Cram, D. J. *J. Am. Chem. Soc.* **1979**, *101*, 4928–4941.

146. Cram, D. J.; Ho, S. P. *J. Am. Chem. Soc.* **1985**, *107*, 2998–3005.

147. Cram, D. J.; Lein, G. M. *J. Am. Chem. Soc.* **1985**, *107*, 3657–3668.

148. Cram, D. J.; deGrandpre, M. P.; Knobler, C. B.; Trueblood, K. N. *J. Am. Chem. Soc.* **1984**, *106*, 3286–3292.

149. Cram, D. J.; Kaneda, T.; Helgeson, R. C.; Brown, S. B.; Knobler, C. B.; Maverick, E.; Trueblood, K. N. *J. Am. Chem. Soc.* **1985**, *107*, 3645–3657.

150. Mitsky, J.; Jaris, L.; Taft, R. W. *J. Am. Chem. Soc.* **1972**, *94*, 3442–3445.

151. Atkins, H. W.; Gilkerson, W. R. *J. Am. Chem. Soc.* **1973**, *91*, 8551–8559.

152. Cram, D. J.; Ho, S. P.; Knobler, C. B.; Maverick, E.; Trueblood, K. N. *J. Am. Chem. Soc.* **1985**, *107*, 2989–2998.

153. Nolte, R. J. M.; Cram, D. J. *J. Am. Chem. Soc.* **1984**, *106*, 1416–1420.

154. Cram, D. J.; Dicker, I. B.; Lauer, M.; Knobler, C. B.; Trueblood, K. N. *J. Am. Chem. Soc.* **1984**, *106*, 7150–7167.

155. Cram, D. J.; Doxsee, K. M.; Feigel, M.; Stewart, K. D.; Canary, J. W.; Knobler, C. B. *J. Am. Chem. Soc.* **1987**, *109*, 3098–3107.

156. Goldberg, I.; Doxsee, K. M. *J. Inclus. Phenom.* **1986,** *4,* 303–322.

157. Stewart, K. D.; Miesch, M.; Knobler, C. B.; Maverick, E.; Cram, D. J. *J. Org. Chem.* **1986,** *51,* 4327–4337.

158. Cram, D. J.; Lein, G. M.; Kaneda, T.; Helgeson, R. C.; Knobler, C. B.; Maverick, E.; Trueblood, K. N. *J. Am. Chem. Soc.* **1981,** *103,* 6228–6232.

159. Kollman, P. A.; Wipff, G.; Singh, U. C. *J. Am. Chem. Soc.* **1985,** *107,* 2212–2219.

160. Helgeson, R. C.; Cram, D. J., unpublished results.

161. Artz, S. P.; Cram, D. J. *J. Am. Chem. Soc.* **1984,** *106,* 2160–2171.

162. Anthonsen, T.; Cram, D. J. *J. Chem. Soc. Chem. Commun.* **1983,** 1414–1416.

163. Cram, D. J. *Angew. Chem. Int. Ed.* **1986,** *25,* 1039–1057.

164. Lein, G. M.; Cram, D. J. *J. Am. Chem. Soc.* **1985,** *107,* 448–455.

165. Chao, Y.; Cram, D. J. *J. Am. Chem. Soc.* **1976,** *98,* 1015–1017.

166. Chao, Y.; Weisman, G. R.; Sogah, G. D. Y.; Cram, D. J. *J. Am. Chem. Soc.* **1979,** *101,* 4948–4958.

167. Matsui, T.; Koga, K. *Tetrahedron Lett.* **1978,** 1115–1118.

168. Lehn, J. –M.; Sirlin, C. *J. Chem. Soc. Chem. Commun.* **1978,** 949–951.

169. Katz, H. E.; Cram, D. J. *J. Am. Chem. Soc.* **1983,** *105,* 135–137.

170. Cram, D. J.; Katz, H. E.; Dicker, I. B. *J. Am. Chem. Soc.* **1984,** *106,* 4987–5000.

171. Cram, D. J. ; Lam, P. Y. S. *Tetrahedron Symposium-in-Print* **1986,** *42,* 1607–1615.

172. Cram, D. J.; Lam, P. Y. S.; Ho, S. P. *J. Am. Chem. Soc.* **1986,** *108,* 839–841.

173. Hubbard, C. D.; Kirsch, J. F. *Biochemistry* **1972,** *11,* 2483–2493.

174. Cram, D. J. *Science (Washington, DC)* **1983,** *219,* 1177–1183.

175. Hogberg, A. G. S. *J. Am. Chem. Soc.* **1980,** *102,* 6046–6050.

176. Moran, J. R.; Karbach, S.; Cram, D. J. *J. Am. Chem. Soc.* **1982,** *104,* 5826–5828.

177. Cram, D. J.; Stewart, K. D.; Goldberg, I.; Trueblood, K. N. *J. Am. Chem. Soc.* **1985,** *107,* 2574–2575.

178. Cram, D. J.; Cram, J. M. In *Selectivity: A Goal for Synthetic Efficiency*; Bartmann, W.; Trost, B. M., Eds., Verlag Chemie: Weinheim, Germany, 1983; Chapter 14, pp 42–64.

179. Helgeson, R. C.; Lauer, M.; Cram, D. J. *J. Chem. Soc. Chem. Commun.* **1982,** 101–103.

180. Helgeson, R. C.; Mazaleyrat, J.–P.; Cram, D. J. *J. Am. Chem. Soc.* **1981,** *103,* 3929–3931.

181. Cram, D. J.; Karbach, S.; Kim, Y. H.; Baczynskyj, L.; Kalleymeyn, G. W. *J. Am. Chem. Soc.* **1985,** *107,* 2575–2576.

182. Cram, D. J.; Hammond, G. S. *Organic Chemistry*; McGraw–Hill: New York, 1959; pp 1–712.

183. Cram, J. M.; Cram, D. J. *The Essence of Organic Chemistry*; Addison–Wesley: Reading, MA, 1978; pp 1–456.

Index

135

Production: Paula M. Befard
Copy Editing and Indexing: Janet S. Dodd
Acquisition: Robin Giroux

Printed and bound by Maple Press, York, PA

*Paper meets minimum requirements of American National Standard
for Information Sciences—Permanence of Paper for Printed Library
Materials, ANSI Z39.48–1984*   ∞